Fu-Sheng Chu

6-25-92

DIGITAL SPEECH PROCESSING
Speech Coding, Synthesis and Recognition

THE KLUWER INTERNATIONAL SERIES
IN ENGINEERING AND COMPUTER SCIENCE

VLSI, COMPUTER ARCHITECTURE AND
DIGITAL SIGNAL PROCESSING

Consulting Editor
Jonathan Allen

Latest Titles

Hardware Design and Simulation in VAL/VHDL, L.M. Augustin, D.C..Luckham,
   B.A.Gennart, Y.Huh, A.G.Stanculescu
      ISBN: 0-7923-9087-3
Subband Image Coding, J. Woods, editor,
      ISBN: 0-7923-9093-8
Low-Noise Wide-Band Amplifiers in Bipolar and CMOS Technologies,
   Z.Y.Chang, W.M.C.Sansen,
      ISBN: 0-7923-9096-2
Iterative Identification and Restoration of Images, R. L.Lagendijk, J. Biemond
      ISBN: 0-7923-9097-0
VLSI Design of Neural Networks, U. Ramacher, U. Ruckert
      ISBN: 0-7923-9127-6
Synchronization Design for Digital Systems, T. H. Meng
      ISBN: 0-7923-9128-4
Hardware Annealing in Analog VLSI Neurocomputing, B. W. Lee, B. J. Sheu
      ISBN: 0-7923-9132-2
Neural Networks and Speech Processing, D. P. Morgan, C.L. Scofield
      ISBN: 0-7923-9144-6
Silicon-on-Insulator Technology: Materials to VLSI, J.P. Colinge
      ISBN: 0-7923-9150-0
Microwave Semiconductor Devices, S. Yngvesson
      ISBN: 0-7923-9156-X
A Survey of High-Level Synthesis Systems, R. A. Walker, R. Camposano
      ISBN: 0-7923-9158-6
Symbolic Analysis for Automated Design of Analog Integrated Circuits,
   G. Gielen, W. Sansen,
      ISBN: 0-7923-9161-6
High-Level VLSI Synthesis, R. Camposano, W. Wolf,
      ISBN: 0-7923-9159-4
*Integrating Functional and Temporal Domains in Logic Design: The False Path
   Problem and its Implications*, P. C. McGeer, R. K. Brayton,
      ISBN: 0-7923-9163-2
Neural Models and Algorithms for Digital Testing, S. T. Chakradhar,
   V. D. Agrawal, M. L. Bushnell,
      ISBN: 0-7923-9165-9
Monte Carlo Device Simulation: Full Band and Beyond, Karl Hess, editor
      ISBN: 0-7923-9172-1
The Design of Communicating Systems: A System Engineering Approach,
   C. J. Koomen
      ISBN: 0-7923-9203-5
Parallel Algorithms and Architectures for DSP Applications,
   M. A. Bayoumi, editor
      ISBN: 0-7923-9209-4

DIGITAL SPEECH PROCESSING
Speech Coding, Synthesis and Recognition

Edited by

A. Nejat Ince
Marmara Research Centre
Gebze-Kocaeli, Turkey

KLUWER ACADEMIC PUBLISHERS
Boston/Dordrecht/London

Distributors for North America:
Kluwer Academic Publishers
101 Philip Drive
Assinippi Park
Norwell, Massachusetts 02061 USA

Distributors for all other countries:
Kluwer Academic Publishers Group
Distribution Centre
Post Office Box 322
3300 AH Dordrecht, THE NETHERLANDS

Library of Congress Cataloging-in-Publication Data

Digital speech processing : speech coding, synthesis, and recognition
  / edited by A. Nejat Ince.
        p.   cm. -- (The Kluwer international series in engineering and
computer science)
        Includes bibliographical references and index.
        ISBN 0-7923-9220-5 (alk. paper)
        1. Speech processing systems.   I. Ince, A. Nejat.   II. Series.
TK7882.S65D54   1992
621.39'9--dc20   91-31404
                                                                      CIP

CONTENTS

Preface...ix

CHAPTER 1:   OVERVIEW OF VOICE COMMUNICATIONS AND
             SPEECH PROCESSING...1

             by A. Nejat Ince

             INTRODUCTION...2
             COMMUNICATIONS NETWORKS....................................4
             OPERATIONAL REQUIREMENTS.................................10
             SPEECH PROCESSING...20
             QUALITY EVALUATION METHODS............................33
             THE SPEECH SIGNAL..36
             CONCLUSIONS..36
             REFERENCES..39

CHAPTER 2:   THE SPEECH SIGNAL...43

             by Melvyn J. Hunt

             INTRODUCTION...44
             THE PRODUCTION OF SPEECH....................................44
             THE PERCEPTION OF SPEECH AND OTHER
                  SOUNDS...54
             SPEECH AS A COMMUNICATIONS SIGNAL.................58
             SPEECH AND WRITING..65
             SUMMARY...70
             REFERENCES..70

CHAPTER 3:   SPEECH CODING...73

             by Allen Gersho

             INTRODUCTION...73
             APPLICATIONS...... 74
             BASICS OF SPEECH CODING...75
             PREDICTIVE QUANTIZATION.......................................75
             LPC VOCODER...79

PITCH PREDICTION...80
ADAPTIVE PREDICTIVE CODING (APC)......................81
VECTOR QUANTIZATION...83
OPEN LOOP VECTOR PREDICTIVE CODING................84
ANALYSIS-BY-SYNTHESIS EXCITATION CODING......85
VECTOR EXCITATION CODING.....................................87
VECTOR SUM EXCITATION CODEBOOKS...................90
CLOSED-LOOP PITCH SYNTHESIS FILTERING.............91
ADAPTIVE POST FILTERING..92
LOW DELAY VXC..94
VXC WITH PHONETIC SEGMENTATION......................96
NONLINEAR PREDICTION OF SPEECH.........................97
CONCLUDING REMARKS..98
REFERENCES...99

CHAPTER 4: **VOICE INTERACTIVE INFORMATION**
SYSTEMS..101

by J. L. Flanagan

INTERACTIVE INFORMATION SYSTEMS....................101
NATURAL VOICE INTERFACES....................................102
AUTODIRECTIVE MICROPHONE SYSTEMS................107
INTEGRATION OF VOICE IN MULTIMEDIA
    SYSTEMS..108
PROJECTIONS FOR DIGITAL SPEECH
    PROCESSING...110

CHAPTER 5: **SPEECH RECOGNITION BASED ON PATTERN**
RECOGNITION APPROACHES...................................111

by Lawrence R. Rabiner

INTRODUCTION..111
THE STATISTICAL PATTERN RECOGNITON
    MODEL..113
RESULTS ON ISOLATED WORD RECOGNITON.........118
CONNECTED WORD RECOGNITION MODEL..............120
CONTINUOUS, LARGE VOCABULARY, SPEECH
    RECOGNITION...123
SUMMARY..124
REFERENCES..125

CHAPTER 6: **QUALITY EVALUATION OF SPEECH PROCESSING SYSTEMS**................................**127**

by Herman J. M. Steeneken

INTRODUCTION...128
SPEECH TRANSMISSION AND CODING
    SYSTEMS..129
SPEECH OUTPUT SYSTEMS.......................144
AUTOMATIC SPEECH RECOGNITION SYSTEMS......147
FINAL REMARKS AND CONCLUSIONS.......................156
REFERENCES...157

CHAPTER 7: **SPEECH PROCESSING STANDARDS**........................**161**

by A. Nejat Ince

  STANDARDS ORGANISATIONS...............................161
WORKING METHODS OF THE CCITT.......................162
CCITT SPEECH PROCESSING STANDARDS...............165
NATO STANDARDISATION ACTIVITIES IN
    SPEECH PROCESSING..............................177
CONCLUSIONS...185
REFERENCES...187

CHAPTER 8: **APPLICATION OF AUDIO/SPEECH RECOGNITION FOR MILITARY REQUIREMENTS**............................**189**

by Edward J. Cupples and Bruno Beek

INTRODUCTION...189
AUDIO SIGNAL ANALYSIS.........................190
VOICE INPUT FOR COMMAND AND CONTROL.......196
MESSAGE SORTING/AUDIO MANIPULATION..........199
AUTOMATIC GISTING......................................202
FUTURE DIRECTION....................................205
REFERENCES...206

SELECTIVE BIBLIOGRAPHY WITH ABSTRACT...**209**

SUBJECT INDEX...**239**

PREFACE

After almost three scores of years of basic and applied research, the field of speech processing is, at present, undergoing a rapid growth in terms of both performance and applications and this is fuelled by the advances being made in the areas of microelectronics, computation and algorithm design.Speech processing relates to three aspects of voice communications:

- Speech Coding and transmission which is mainly concerned with man-to-man voice communication.
- Speech Synthesis which deals with machine-to-man communication.
- Speech Recognition which is related to man-to-machine communication.

Widespread application and use of low-bit rate voice codecs, synthesizers and recognizers which are all speech processing products requires ideally internationally accepted quality assessment and evaluation methods as well as speech processing standards so that they may be interconnected and used independently of their designers and manufacturers without costly interfaces.

This book presents, in a tutorial manner, both fundamental and applied aspects of the above topics which have been prepared by well-known specialists in their respective areas. The book is based on lectures which were sponsored by AGARD/NATO and delivered by the authors, in several NATO countries, to audiences consisting mainly of academic and industrial R&D engineers and physicists as well as civil and military C3I systems planners and designers.

The book starts with a chapter which discusses first the use of voice for civil and military communications and considers its advantages and disadvantages including the effects of environmental factors such as acoustic and electrical noise and interference and propagation. The structure of the existing NATO communications network is then outlined as an example and the evolving Integrated Services Digital Network (ISDN) concept is briefly reviewed to show how they meet the present and future requirements. It is concluded that speech coding at low-bit rates is a growing need for transmitting speech messages with a high level of security and reliability over capacity limited channels and for memory-efficient systems for voice storage, voice response, and voice mail etc. Furthermore it is pointed out that the low-bit rate speech coding can ease the transition to shared channels for voice and data and can readily adopt voice messages for packet switching. The speech processing techniques and systems are then briefly outlined as an introduction to the succeeding sections.

Chapter 2 of the book provides a non-mathematical introduction to the speech signal itself. The production of speech is first described, including a survey of the categories into which speech sounds are grouped. This is followed by an account of some properties of human perception of sounds in general and of speech in particular. Speech is then compared with other signals. It is argued that it is more complex than artificial message bearing signals, and that unlike such signals speech contains no easily identified context-independent units that can be used in bottom-up decoding. Words and phonemes are examined, and phonemes are shown to have no simple manifestation in the acoustic signal. Speech communication is presented as an interactive process, in which the listener actively reconstructs the message from a combination of acoustic cues and prior knowledge, and the speaker takes the listener's capacities into account in deciding how much acoustic information to provide. The final section compares speech and text, arguing that our cultural emphasis on written communication causes us to project properties of text onto speech and that there are large differences between the styles of language appropriate for the two modes of communication. These differences are often ignored, with unfortunate results.

Chapter 3 deals with the fundamental subject of speech coding and compression. Recent advances in tecnhniques and algorithms for speech coding now permit high quality voice reproduction at remarkably low bit rates. The advent of powerful single-chip signal processors has made it cost effective to implement these new and sophisticated speech coding algorithms for many important applications in voice communication and storage. This chapter reviews some of the main ideas underlying the algorithms of major interest today. The concept of removing redundancy by linear prediction is reviewed, first in the context of predictive quantization or DPCM, then linear predictive coding, adaptive predictive coding, and vector quantization are discussed. The concepts of excitation coding via analysis-by-synthesis, vector sum excitation codebooks, and adaptive postfiltering are explained. The main idea of Vector Excitation Coding (VXC) or Code Excited Linear Prediction (CELP) are presented. Finally low-delay VXC coding and phonetic segmentation for VXC are described. This section is concluded with the observation that mobile communications and the emerging wide scale cordless portable telephones will incresingly stress the limited radio spectrum that is already pushing researchers to provide lower bit-rate and higher quality speech coding with lower power consumption, increasingly miniaturized technology, and lower cost. The insatiable need for humans to communicate with one another will continue to drive speech coding research for years to come.

In Chapter 4 an overview of voice interactive information systems is given aimed at highlighting recent advances, current areas of research, and key issues for which new fundamental understanding of speech is needed. This chapter also covers the subject of speech synthesis where the principal objective is to produce natural quality synthetic speech from unrestricted text input. Useful applications of speech synthesis include announcement machines (e.g. weather, time) computer answer back (voice messages, prompts), information retrieval from databases (stock price quotations, bank balances), reading aids for the blind, and speaking aids for the vocally handicapped. There are two basic methods of synthesizing speech which are described in this chapter: The first and easiest method of providing voice output for machines is to create speech messages by concatenation of prerecorded and digitally stored words, phrases, and sentences spoken by a human. However, these stored-speech systems are not flexible enough to convert unrestricted printed text-to-speech. In the text-to speech systems the incoming text including dates, times, abbreviations, formulas and wide variety of punctuation marks are accepted and converted into a speakable form. The text is translated into a phonetic transcription, using a large pronouncing dictionary supplemented by appropriate letter-to-sound rules. Both of these methods are compared in this chapter in terms of quality (naturalness), the size of the vocabulary, and the cost which is mainly determined by the complexity of the system.

Probably the most intractable of all the speech processing techniques is speech recognition where the ultimate objective is to produce a machine which would understand conversational speech with unrestricted vocabulary, from essentially any talker. Algorithms for speech recongnition can be characterized broadly as pattern recognition approaches and acoustic phonetic approaches. To date, the greatest degree of success in speech recognition has been obtained using pattern recognition paradigms. It is for this reason that Chapter 5 is concerned primatily with this technique. A pattern recognition model used for speech recognation is first described.The input speech signal is analysed (based on some paremetric model) to give the test pattern which is compared to a prestored set of reference patterns using a pattern classifier.The pattern similarity scores are then sent to a decision algorithm which, based upon the syntax and/or semantics of the task chooses the best transcription of the input speech. This model is shown to work well in practice and is therefore used in the remainder of the chapter to tackle the problems of isolated word (or discrete utterences) recognition, connected word recognition, and continuous speech recognition. It is shown that our understanding (and consequently the resulting recognizer performance) is best for the simplest recognition tasks and is considerably less well developed for large scale recognition systems. This chapter concludes with the observation that the performance of current systems is barely acceptable for large vocabulary systems, even with isolated word inputs, speaker training, and favourable talking environment and that almost every aspect of continuous speech recognition, from training to systems implementation, represents a challenge in performance, reliability, and robustness.

There are different, but as yet universally not standardized, methods (subjective and objective) to measure the "goodness" or "quality" of speech processing systems in a formal manner.The methods are divided into three groups: Subjective and objective assesment for speech coding/transmission and speech output systems (synthesizers) and thirdly assesment methods for automatic speech recognition systems.These are discussed in Chapter 6. The evaluation of the first two systems is done in terms of intelligibility measures. The evaluation of speech recognizers is shown to require a different approach as the recognition rate normally depends on recognizer-specific parameters and external factors. Howewer, more generally applicable evaluation methods such as predictive methods are also becoming available. For military applications, the test methods used include the effects of the environmental conditions such as noise level, acceleration, stress, mask microphones which are all referred to in this chapter. Results of the assessment methods as well as case studies are also given for each of the three speech systems. It is emphasised that evaluation techniques are crucial to the satiscfactory deployment of speech processing equipments in real applications.

Chapter 7 deals with international (global, regional CEPT and NATO) speech processing standards the purpose of which is "to achieve the necessary or desired degree of uniformity in design or operation to permit systems to function beneficially for both providers and users" i.e. ,interoperable systems without complex and expensive interfaces. The organization, working methods of CCITT (The International Telegraph and Telephone Consultative Committe) and of NATO as well as the procedures they use for speech processing standards are explained   including test methods and conditions. The speech processing standards promulgated by CCITT within the context of ISDN are described in terms of encoding algorithms and codec design and their performance for voice and voice-band data are discussed as a function of transmission impairements and tandem encoding. These standards are related to the so-called "low-bit-rate-voice" (LBRV) which aim at overcoming, in the short-to-medium terms (before the widespread use of the emerging optical fibre) the economic weakness of 64 kb/s PCM in satellite and long-haul terrestrial links and copper subscriber loops and also to "high-fidelity voice" (HFV) with bandwidth up to 7 kHZ for applications such as loudspeaker telephones, teleconferencing and commentary channels for broadcasting. Other CCITT activities for future standards are also discussed in this chapter which relate to Land Digital Mobile Radio (DMR), SCPC satellite links with low C/N, Digital Circuit Multiplication Equipment (DCME) and packetized speech for the narrow-band and the evolving wide-band ISDN with 'asychronous transfer mode'.

This chapter concludes with a description of two NATO speech processing standards in terms of algorithms, design and test procedurs; 2.4 kb/s Linear Predictive Coder (LPC) and 4.8 kb/s Code Excited Predictive Coder (CELP), both for secure voice use on 3 kHz analog land lines and on High-Frequency radio channels. A third NATO draft standardization agreement is also mentioned, for the sake of completeness, which concerns A/D conversion of voice signals using 16 kb/s Delta Modulation and Syllabic Companding (CVSD).

The last chapter of the book, Chapter 8, entittled "Audio/Speech Recognition for Military Applications", examines some recent applications of ASR technology which complement the several civil applications mentioned in the previous chapters. Four major categories of applications are discussed which are being pursued at the Rome Air Development Center (RADC) to satisfy the US Air Force requirements for modern communication stations and the FORECAST II Battle Management and Super Cockpit Programs:

- Speech Enhancement Technology to improve the quality, readability and intelligibility of speech signals that are masked and interfered with by communication channel noise so that humans may listen and understand and machines may process the signals received.
- Voice input for Command and Control including automatic speaker verification to verify the identity of individuals seeking access to restricted areas and systems.
- Message Sorting by Voice which tries to automate part of listening to radio broadcasts. A Speaker Authentication System (SAS) is outlined in this section which uses two techniques, a multiple parameter algorithm employing the Mahalanobis metric and an identification technique based on a continuous speech recognition algorithm.
- Speech Understanding and Natural Language Processing for the DOD Gister Program which aims at automatically 'gist'ing voice traffic for the updating of databases to produce in-time reports.

Chapter 8 Concludes with information on future direction of work which is being carried out at RADC including the development of a VHSIC speech processor that can provide the processing power to support multiple speech functions and channels.

The book ends with an extensive bibliography with abstracts which has been prepared with the kind assisstance of the Scientific and Technical Information Division of the U.S. National Aeronautics and Space Administration (NASA), Washington, D.C.

Prof. A. Nejat INCE

CHAPTER 1

OVERVIEW OF
VOICE COMMUNICATIONS
AND SPEECH PROCESSING

A. Nejat Ince *

Istanbul Technical University
Ayazaga Campus
Istanbul
Turkey

There are three uses of speech : the first is to express ideas , the second is to conceal ideas, and the third is to conceal the lack of ideas!

ABSTRACT

This chapter discusses the use of voice for civil and military communications and outlines possible operational requirements including environmental factors and the effects of propagational factors and electronic warfare. Structures of the existing NATO communications network (taken as an example of military networks) and the evolving Integrated Services Digital Network (ISDN) are reviewed to show how they meet the requirements postulated.

It is concluded that speech coding at low-bit rates is a growing need for transmitting speech messages with a high level of security and reliability over low data-rate channels and for memory-efficient systems for voice storage, voice response, and voice mail etc. Furthermore it is pointed out that the low-bit rate voice coding can ease the transition to shared channels for voice and data and can readily adopt voice messages for packet switching.

The speech processing techiques and systems are then outlined as an introduction to the lectures of this series in terms of:

- The character of the speech signal, its generation and perception
- speech coding which is mainly concerned with man-to-man voice communication
- speech synthesis which deals with machine-to-man communication
- speech recognation which is related to man-to-machine communication
- quality assessment of speech system and standards

---

* Prof Ince is Director of Marmara Research Center of The Turkish Science Council TUBITAK.

1. INTRODUCTION

Although there are many shades of opinion, communication is broadly defined to be the establishment of social unit from individuals, by the use of language or signs [1]. When we communicate, one with another, we make sounds with our vocal organs, or scribe different shapes of ink mark on paper (or some other medium), or gesticulate in various patterned ways; such physical signs or signals have the ability to change thoughts and behaviour-they are the medium of communication. Telecommunications engineers have as their business the extension of the distance over which the communication process normally takes place by transmitting such signals while preserving their forms in such systems as telephones, telegraphs, facsimile, video.

It must be noted here that the "social unit" in a multinational environment such as NATO is multilingual and multinational with all that these imply in exchanging or sharing information which make it different from a more homogenious national environment. Greater care must therefore be exercised in using national results relating to speech input/output systems. One feels instinctively that communications in such a case would somewhat be more difficult, complex, less accurate and longer, thus making written communications more important.

There are two distinct classes of signal. There are signals in time such as speech or music; and there are signals in space, like print, stone inscription, punched cards, and pictures. Out of all these communication forms, "speech" is perhaps, the most "natural" mode by which human beings communicate with each other. There are also good reasons for people wishing to use speech to communicate with machines. It must, however, be pointed out that there is not much empirical evidence to show the value of speech over other modes of communication.

In a recent study carried out by the author [2] it was established that in a tri-service strategic C3I environment about half of the total traffic in Erlangs was for voice and the rest was approximately equally divided between data and message traffic. In an information theoretic sense, however, the bulk of communication was carried by the message handling system. About 70% of the traffic was for air operations. It is, however, expected that these proportions will change with time in favour of the data traffic. The traffic situation is, of course, very different in the civil network where, at least in the foreseeable future, voice service will continue to predominate all others. It must be stated however, that the main reason for the preponderance of message traffic in military networks today is the requirement of "recording" information in a secure and easily accessible way and ability to coordinate and disseminate it.

Notwithstanding the above, an experiment carried out at Johns Hopkins University [3] showed that teams of people interacting together to solve problems solved them much faster using voice than any other mode of communication. There are other studies which indicate that voice provides advantages over other means of communication for certain applications. There is no doubt that the main reason for the preference of voice, at least for certain applications, stems from it being "natural", not requiring any special training to learn, and freeing the hands and eyes for other tasks.

The features of speech communications that are disadvantageous relate to the difficulty of keeping permanent secure records, interference caused by competing environmental acoustic noise, physical/psychological changes in the speaker causing changes in speech characteristics or disabilities of speaking/hearing and finally its serial and informal nature leading to slower information transfer and/or information access. It must be pointed out however that some of the disadvantages of speech communication are dependent on the state of technology and can therefore change with time and application.

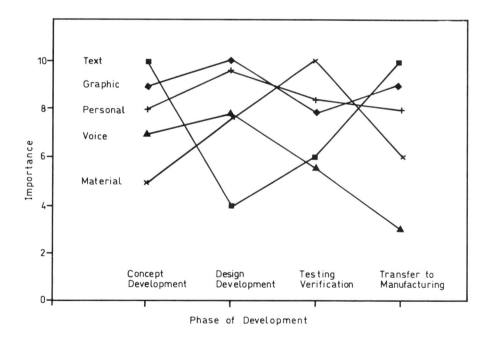

Fig.1. Encoding of Communications in an Engineering Development

Fig.1 shows how the importance of the communication mode changes with the phases of an engineering project [4]. The importance of text dominates at the beginning and end of an engineering development process. In the middle of the process, other forms of communication modes rise and fall in importance, due to the specialised design and implementation methods of engineering. Graphics maintains its importance throughout the process.

From the example above it is not too difficult to see a certain degree of resemblence between the modes of communication required for an engineering development project and those for command and control; all modes are required in general with preference given to some depending on application and the development of technologies and operational concepts.

2. COMMUNICATIONS NETWORKS

Because of the preponderance of traffic related to Air Operations in military networks we shall now take a brief look at the type of communications these operations require and the type of environment in which they are to work.

Air operations involve both fixed and mobile platforms (land, sea and air) and communications that are required to interconnect them consist of:
- A switched terrestrial network
- Air/ground communications and
- Intra-aircraft (cockpit) communications.

These communications are used to support:
- the management of offensive air operations
- the management of defensive aircraft
- regional, sub-regional air defence control systems.

In addition there are also dedicated communications employed for sustained surveillance, navigation and IFF.

The main air warfare missions and associated ranges together with the types of communications required are given in Fig 2. These communications are currently provided in NATO by a combination of international and national networks using both terrestrial and satellite links together with VHF/UHF ground/air, air/air and HF radio communications to and between tactical/strategic aircraft (Fig. 3).

The terrestrial transmission systems used today provide nominally 4 kHz analogue circuits even though the NATO SATCOM systems is totaly digitised and some national systems (PTT and military) use digital transmission links. NATO also owns and operates automatically switched voice and telegraph networks. It is to be noted that a significant portion of the traffic that flows in the common-user

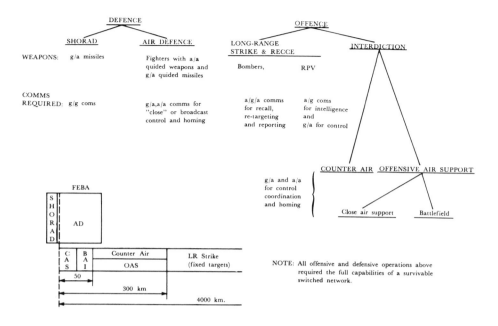

Fig.2. Air Warfare Missions and Range

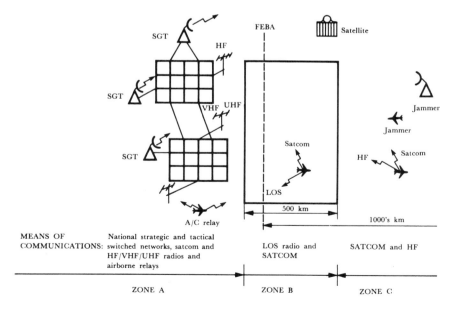

Fig.3. Communications Zones and Means

network is related to air operations. As far as UHF/VHF and HF radios are concerned, they provide analogue voice and data except for JTIDS (The Joint Tactical Information Distribution System) which is totally digital and is currently available for the NATO AEW program. The NATO communications systems carry some circuits which are cryptographically secure end-to-end and there are some links and circuits carried by SATCOM and JTIDS which are protected also against jamming.

2.1 Integrated Services Digital Network (ISDN)

NATO decided in 1984 that most of their terrestrial communications requirements would be met in the future by the strategic military communications networks that are today being designed and some being implemented by the Member Countries. All these networks largely follow the CCITT IDN/ISDN standards and recommendations and adopt the International Standards Organisation's (ISO) Open System Interconnection Reference Model (OSI/RM). These digital common-user grid networks provide mission related, situation oriented, low-delay "teleservices" such as plain/secure voice, facsimile and non-interactive and interactive data communications. These are enhanced by "supplementary services" such as priority and pre-emption, secure/non-secure line warning as well as closed-user groups, call forwarding and others. The switching subsystem supports three types of connection methodology namely, semi-switched connections, circuit-switched connections, and packet/message switched connections. The circuit switching technique used is the byte-oriented, synchronous, time-division-multiplexed (TDM) switching in accordance with CCITT standards. The basic channels are connected through the network as transparent and isochronous circuit of 64 kb/s or nx64 kb/s where n is typically 32. Possible uses of the 64 kb/s unrestricted circuits are shown in Fig 4.

The basic channel structure used in ISDN has T and S reference points and consists of two B channels at 64 kb/s and one D channel at 16 kb/s. One or both of the B channels may not be supported beyond the interface. The B channel is a pure digital facility (that is, it can be used as circuit-switched, packet-switched, or as a non-switched/nailed facility), while the D channel can be used for signalling, telemetry, and packetswitched data. The basic access allows the alternate or simultaneous use of a number of terminals. These terminals could deal with different services and could be of different types. The basic architectural model of an ISDN is shown Fig. 5.

The primary rate B-channel structure is composed of 23 B or 30 B channels (depending on the national digital hierarchy primary rate, that is, 1544 or 2048 kb/s and one D channel at 64 kb/s. PABX connection to the T reference point can use (depending on its size) multiple basic channel structure accesses, a primary rate

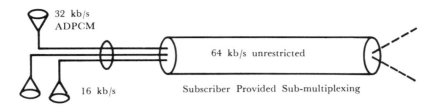

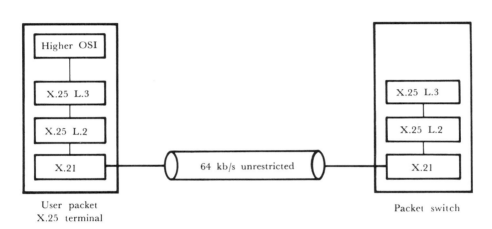

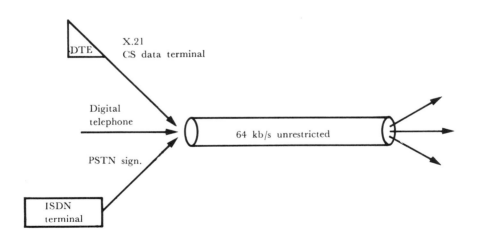

Fig.4. Scenarios for the Use of the 64 kb/s Unrestricted
Circuit Mode Channel

8

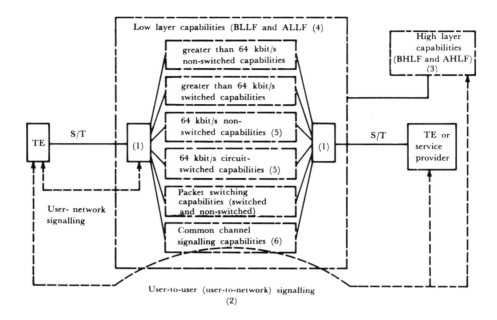

Fig.5. Basic Architectural Model of an ISDN

Note 1 - The ISDN local functional capabilities corresponds to functions provided by a local exchange and possibly other equipments such as electronic cross connect equipments muldexes, etc.

Note 2 - User-to-user signalling needs further study.

Note 3 - These functions may either be implemented within ISDN or be provided by separate networks.

Note 4 - In certain national situations, ALLF may also be implemented outside the ISDN, in special nodes or in certain catagories of terminals.

Note 5 - Circuit switching and non-switched functional capabilities at rates less than 64 kbit/s are for further study.

Note 6 - For signalling between international ISDNs, CCITT No. 7 shall be used.

B-channel structure, or one more primary rate transmission systems with a common D channels. The primary rate H-channel interface structures are composed of Ho channels (384 kb/s) with or without a D channel, or an H1 channel (1536 kb/s). H channels can be used for high-fidelity sound, high-speed facsimile, high-speed data, and video. Primary rate mixed Ho and B-channel structures are also possible. Subrate channel structures are composed of less than 64 kb/s channels and are rate adapted and multiplexed into B channels.

Future evolution of the ISDN will likely include the switching of broadband services at bit rates greater than 64 kb/s , at the primary rate, as well as switching at bit rates lower than 64 kb/s which are made possible by the end-to-end digital connectivity. Table I shows some typical service requirements for civil and also for military applications.

In the ISDN environmet, the use of common channel signalling networks significantly reduces the call setup and disconnect times; use of Digital Speech Interpolation (DSI) can enhance the transmission efficiency on a cost-driven basis. Packet switching [5,6] which allocates bandwith on a dynamic basis, has become the preferred technique for data communications. In addition to utilising the bandwith more efficiently, packet switching permits protocol conversion, error control, and achieves fast response times neede for interactive data communications.

Table I: Some Service Requirements

Service	Bandwith Requirements	ISDN Channel Type B	D	Circuit Switched	Packet Switched	Channel Switched	Overlay
*Telephone	8,16,32,64kb/s	x	x				
*Interactive Data Communications	4.8-64 kb/s	x		x	x		
*Electronic Mail	4.8-64 kb/s	x			x		
*Bulk Data Transfer	4.8-64 kb/s	x		x			
*Facsimile/ Graphics	4.8-64 kb/s	x		x			
*Slow Scan/ Freeze Frame TV	56-64 kb/s	x		x			
*Compressed Video Conference	1.5-2 Mb/s (Primary rate)					x	x

Looking ahead into the future both for military and civil applications, we see good prospects for the integration of voice and data traffic. Investigation of different techniques permitting integration of voice and data traffic in one network has been a subject of ongoing research for more than a decade. These techniques include hybrid switching [7], burst switching [8], and packet switching for speech and data [9]. A common objective of all these techniques is to improve efficiency of speech connections in comparison with the circuit-switched network, with minimal degradation to speech quality as a result of clipping and message delay.

Hybrid switching can achieve acceptable voice message delays. However, lower transmission efficiency and higher complexity than packet-switching concepts render it unattractive for application in public switched networks.

Burst switching achieves high transmission efficiencies and low voice message delays. It is an attractive concept, but high costs associated with the development of a new family of switching systems and the lack of evolutionary migration paths for implimentation make it unsuitable for public networks.

The attraction of speech packet communications [9] lies in the relative simplicity of packet-switching concepts, and the fact that computer systems for data packet switching can be adopted for speech packet comumnications. While existing protocols for packet data communications such as X.25 are not suitable for achieving small fixed delays necessary in speech packet communications, significant progress has been made in developing new protocols under the sponsorship of the Defence Advanced Projects Agency (DARPA) [10,11] and the Defence Communication Agency (DCA). While still in a developmental stage, speech packetisation increasingly appears to be the prime contender for future voice/data integration in common-user networks.

Another speculative impetus for speech packet communications lies in the potential for voice recognition and direct speech input to program, command, and control the operation of artificial intelligence machines. Speech packet communications are ideally suited for such applications.

3. OPERATIONAL REQUIREMENTS

The requirements for air operations are subsumed in the total requirement for the switched networks. The network must be dimensioned to meet the needs of non-mobile military traffic securely, reliably, survivably, and with no operationally significant delays, so as to preserve the radio-frequency spectrum for mobile and broadcast applications, including possibly the restoration or reconfiguration of the static network and/or rerouting of traffic following battle damage. The survivability of the communications must (at least) match that of the war headquarters and weapon sites which it integrates and serves. Operational procedures must be developed to maintain essential operation, even when the capacity of this network has been seriously reduced by battle damage. Survivability of connectivity is however of paramount importance.

The satellite network must similarly be dimensioned to meet the joint requirements of its total user community which comprises primarily those difficult to access otherwise because of:

a) Long range (and relatively large data-rate) requirements

b) mobility,

c) multi-access requirements.

Its security and ECM resistance must be assured and its potential any-to-any and any-to-all capability must be made available for flexible explotation by the user.

Security and ECM resistance are equally required for the various tail links.

Air-ground and ground-air links for close fighter control cannot tolerate delays of more than a fraction of a second when they are part of a close-control loop. The true data-rate in information-theory terms, is not more than 100 bits. It is essential that the interface to the pilot will be user-friendly, and this should normally include (possibly synthesized) spoken messages, in order to keep the pilot's eyes free for his primary duties. Immunity to even short-term disruption by ECM is essential. The air-ground capacity required is marginally smaller than that for ground-air.

Broadcast Control can accept slightly longer delays, but it involves a more varied type of data and may involve a somewhat larger total data rate; it may also require more air-ground traffic. The need for communications with close-support strike aircraft, in a confused and rapidly changing battle situation are similar to those for fighters, but with increased flexibility and capacity in the air-ground direction.

Long-range deep-penetration missions must be accessible to relatively few and short re-targeting and recall messages. In principle, the data rates need to exceed a few bits per second, and delays of possibly several minutes could be tolerated if necessary. In the reply direction acknowledgements and reports of survival or otherwise, and of success or failure of a strike mission are equally undemanding in terms of communications capacity. Any reconnaisance reports from long range could also tolerate a delay of a few minutes if necessary, but even with data reduction, reconnaissance reports (from any range) can benefit from the widest bandwidth which can be provided with the technology available. For long-range missions, low probability of intercept would also be highly desirable.

If the technology dictates a sharp division in capability and/or solution between operations:

    a) within line-of-sight from the ground behind the FEBA,

    b) within line-of-sight from the air behind the FEBA,

    c) beyond line-of-sight from the FEBA,

good, but distinct, solutions to these three scenarios can be accepted.

The operational requirements outlined above do certainly imply, in addition to graphic and data communications, the use of voice. Intelligibility is the most important parameter with "speaker recognition" aiding "authentication" being also required although its value in a multinational environment may be questioned.

3.1 Satellite and HF Channels and Electronic Warfare

We must now turn our attention to the restrictions that propagation conditions and jamming impose on the capacity of HF and satellite channels that are to be used to support long-distance communications to and from the mobile platforms.

HF is in use as a primary means of communication between aircraft and the ground over distances beyond the line-of-sight (LOS) for naval communications, ship-ship, ship-shore and ship-air. Its principal advantage is that it provides connectivity at low cost, so that it will continue to be used in a variety of roles:

    - on large aircraft (e.g. bombers, AEW) as back-up to SATCOM to increase
      the cost to the enmy of ECM and to provide medium redundancy
    - on small aircraft (e.g. fighters, helicopters) which are not provided with
      SAT-COM
    - for a wide variety of naval communications.

Present systems are perceived to have a number of weaknesses in addition to the inherent dispersive nature of the channel itself. However techniques have been proposed which could alleviate or eliminate those weaknessess [35]. It is believed that providing such techniques are employed, HF will continue to provide connectivity at low cost even in the more difficult jamming environment to be expected in the future. It must be recognized however that high bit rates are not considered to be achievable - what is offered is a bit rate in the order of 2.4 kbits/s under favourable conditions, degrading to about 100 bits/s under severe jamming conditions. It must also be recognized that 99.99% availability is not achievable, since even if the effects of interference etc. (which provide the principal limitation of present systems) are overcome, there are residual effects such as disturbance of the medium by various natural causes, and possible nuclear explosions in the atmosphere, which will make it extremely difficult, if not impossible, to increase availability above say 95%. Satellite communications to be used both for the switched networks as well as for mobile users is expected to consist of multiple satellites operating both in the 8/7 GHz SHF band also in the 44, 30/20 GHz EHF band.

In addition to geosynchronous equatorial orbits, inclined non-circular (molnya) orbits are expected to be utilised to provide SATCOM coverage extending up to the polar regions.

These satellites will use multi-beam receiver antennas with nulling capability and multi-beam transmit antennas and processing transponders as a measure for countering the ECM threat [36].

Some of these satellites will be owned and operated by NATO while others will be owned and operated by various NATO nations. The SATCOM capacity offered by these NATO and national assets will be exchanged under various Memoranda of Understanding (similar to current practice) to increase the survivability of NATO and national military common-user networks; this may require interoperability between NATO and national systems.

The main advantages of EHF SATCOMs for communication using small terminals on mobile platforms, as compared to the presently used SHF and UHF SATCOMs are increased anti-jam (AJ) capability, improvement in covertness of communications and increased immunity to the disturbing effects caused in the propagation path by high altitude nuclear detonations.

One geostationary satellite situated over the east Atlantic can provide sufficient coverage for communication among terminals within the NATO ACE (and also Atlantic) region. To provide coverage at latitudes above approximately 65° (especially if communications from the polar regions are required), a constellation of satellites utilizing inclined noncircular orbits will be required. Inter-satellite links may be used to provide connectivity between users accessing different satellites.

The EHF satellites serving the airborne users are expected to use the 44 GHz uplink and 20 GHz downlink frequency bands with the satellite bandwith available in the uplink and the downlink directions being 2 GHz and 1 GHz respectively. Frequency hopping is expected to be used as the spread-spectrum AJ modulation technique so as to fully exploit the available transmission bandwidths (and also to minimize the disturbances from high altitude nuclear explosions)[36].

On board processing involving dehopping/rehopping or dehopping/ demodulation/ remodulation/ rehopping techniques are expected to be utilized in these satellites. Such a processing transponder will provide AJ performance improvement superior to that which can be provided by a conventional non-processing transponder. Furthermore, such a processing transponder will transform the available 2 GHz uplink bandwidth into a 1 GHz downlink bandwidth hence permitting the full utilization of the wider spreading bandwidth available in the uplink direction.

It is assumed that these satellites will use multi-beam receive antennas with adaptive spatial nulling capability and multiple spotbeam transmit antennas for increased jamming resistance.

The critical direction in a SATCOM link to an aircraft will be transmission from the aircraft in the home base direction because the aircraft SATCOM terminal has a small transmit EIRP.

It can be shown that an aircraft having an EHF SATCOM terminal with 60 dBW EIRP can support a data rate of approximately 600 bps under a postulated maximum level of uplink jamming of say 125 dBW EIRP. This traffic capacity assumes the use of a processing satellite with 2 GHz spead spectrum (hopping) bandwidth and a nulling Satellite receive antenna with 35 dB nulling in the jammer direction. The method of calculating the jammed traffic capacity is given in the Annex. It should be noted that the calculated traffic capacity is not a function of the type of orbit used by the satellite.

Downlink jamming of the aircraft receiving terminal is considered a lesser threat since the use of spread spectrum techniques and highly directional receiving antennas with low sidelobes would have to be in line-of-sight to the jammer and would require the jammer to use a directional antenna and this needs to be repeated for each aircraft.

As can be seen in Fig 3, relay aircraft may be used to provide unrestricted communications to aircraft or missiles up to about 200 km beyond the FEBA. As for HF and satellite links, relayed links to the aircraft would also be vulnerable to ground-and air-borne jammers. The achievable maximum range ratio R (TX to RX distance divided by jammer to RX distance) for a given data rate and threat level is obtained from:

$$R^2 = (J/S).(P_s/P_j)$$

where an ultimate anti-jam margin is given by $J/S = (200/B)(1/3)$. This assumes a spread bandwith of 200 MHz and B is the data rate in Mb/s. This relation is plotted in Fig 6. Under a pessimistic assumption: $R = 10:1$ in favour of the enemy and $P_s = 100$ W and $P_j = 100$ kW gives a maximum data rate of about 200 bits/s.

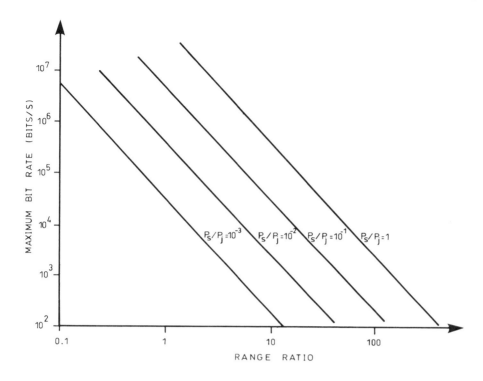

Fig.6. Probable Maximum Bit Rates as a Function of Range
Ratio. Signal-to-Jammer Power Ratio as a Parameter
for a 200 MHz Spectrum Bandwith

3.2 Cockpit Engineering [12]

The basic piloting functions are the following:
- flying (control of aircraft manoeuvres)
- navigation (location and guidance)
- communications (voice and data link)
- utilities management
- mission management

Decisions to be made by the pilot related to the above tasks are crucially
influenced by how information is obtained, and displayed   and   how

communications are processed and handled. There is also the problem of language between the machine and the man. Even when the machine is as learned as the pilot, it will not always know what part of its knowledge is to be transmitted to the pilot or how to optimally transmit it.

It is generally accepted that in many current military aircraft, particularly single-crew aircraft, pilot workload is excessive and can be a limit to the capability of the aircraft as an operational weapon system. Advances in on-board avionics systems have the potential for generating more information, and considerable care will be required in optimising the man-system interface in order that the human pilot capability (which will be essentially unchanged) is not a major constraint on overall system performance.

Ideally the man and aircraft systems would together be designed as a total system. This concept is constrained by some special features of the man which include his performance variability (from man-to-man and from day-to-day) the methods required to load information into him, and his particular input/output channels. At the present time the man has some important capabilities which, in the short term, are unlikely to be attainable with machines. These include:

- Complex pattern recognition

- Assessment and decision-making concerning complex situations

- Intuitive judgement.

Although computers currently excel in analysis and numerical computation, their capabilities in the field of artificial intelligence are developing to the point at which the man's capabilities in complex assessment and decision-making may be overtaken. The implications for the design of man-machine interfaces have not yet been explored, and could raise some important and fundamental new issues.

Another difference between man and machine is in integrity and failure mechanisms. For the forseable future, man is likely to have a unique capability to combine extremely high integrity with complex high-bandwidth operation. The integrity implications of artificial intelligence will certainly require much study. At the same time it must be recognised that the demands on pilots of modern aircraft are such that accidents happen far too frequently, and it should be an aim of overall system design to reduce the frequency of human failure. Better simulation and briefing prior to flying the aircraft may be an important development which will arise from new electronic techniques.

Current advances in control display technology may be projected into the future and we may predict that, by 2005-2010 we will be able to operationally field advanced devices within the cosntraints of tactical aircraft size, weight, and cost.

In this time period digital processing is expected to be orders of magnitude less costly, in terms of size, weight and dollars, than current equipment. In addition we may confidently expect that AI languages and progamming aids will make it much easier to generate complex computer programs that will be able to cost-effectively solve problems that currently require human intelligence. These advances will result in:

- Head-up and eyes-out panoramic displays with large fields of view, high resolution, color, and if desired, enhanced stereo depth cues.
- Ability to synthesize "real world" imagery and pictorial tactical situation displays that recreate clear day visual perception under night and adverse weather conditions.
- High quality voice synthesis and robust voice recognition.
- Natural control of sensors, weapons system, aircraft flight, and display modes based on head and eye position, finger position, and other "body language" modalities.
- Great simplification of tasks that require transfer of complex tactical situation information from the system to the aircrew, and rapid application of the aircrew's superior cognitive powers to management of the weapon system.

Around the year 2000 aircraft displays are not only windows on the status of flight but are vital in the decision making process during certain stages of the mission. Especially during those phases with a high pilot workload, the mission and aircraft data must be formatted and displayed in such a way that the quality and rate of information to be extracted by the pilot is sufficient to arrive at major complex decisions within 2 to 5 seconds without exceeding the pilot's peak workload capacity. In some cases this also implies that the pilot must delegate some of the lower priority (but still important) decisions to an automated device without risk of conflict. The displays should also enable him to evaluate such risks.

3.3 The Use of Voice Systems in the Cockpit

Visual signals are spatially confined; one needs to direct the field-of-view, moreover, in high workload phases of the mission, attention can be focussed on some types of visual information such that other information which suddenly becomes important can be "overloaded". Also the amount of information may saturate the visual channel capacity. Aural signals have the advantage of being absorbed independently of visual engagement, while man's information acquisition capacity is increased by using the two channels simultaneously. Motoric skills are hardly affected by speaking. For information being sent from aircraft to other humans on the ground and in the air, speech is a natural and efficient technique which has been used for many years. Until now the process of aural communication between aircrew and systems has been usually restricted to a limited range of warning signals generated by the systems.

Digital voice synthesis devices are now widely available and have many commercial applications (see section 4.3). Technically there appear to be no problems in using them in aircraft to transfer data from aircraft systems to aircrew, the real difficulty being in identifying the types of message which are best suited to this technique. Warning messages currently appear to be a particularly useful application, though these will probably need to be reinforced by visual warnings as aircrew can totally miss aural warnings under some conditions. Feedback of simple numerical data is also being considered.

One of the main disadvantages of aural signals is that the intelligibility is greatly impaired by noise in the cockpit; this is true both ways. However, the understanding of the mechanisms of speech synthesis and speech recognition has reached the point where voice systems in the cockpit can be considered. Although electronic voice recognition in the laboratory reaches scores of 96 to 98% (comparable with keyboard inputs) the vocabulary is still very limited and recognition tends to be personalised. But the prospect of logic manipulation in AI techniques can greatly improve the situation to depersonalise recognition in noisy environments. Actual data on such improvements are difficult to obtain. These would also depend on how much redundancy is used in both syntax and semantics. Furthermore a coding "language" is to be preferred just as in conventional aircraft radio communication, to prevent the system responding to unvoluntarily uttered (emotional) exclamations.

Several commercial voice recognition equipments are currently available on the open market, but many of these have not been designed for airborne application and considerable development will be needed before they can be regarded as usable equipments for combat aircraft. Simulator and airborne trials in a number of countries using this early equipment have identified the following as key areas in which further investigation/improvement is required:

a) Size of vocabulary. At present this is very limited, but recognition performance is generally inversely related to vocabulary size.

b) Background noise/distortion. The cockpit environment is frequently very poor, and the oxygen mask and microphone are far from ideal.

c) Necessity for pre-loading voice signatures. Current systems have to be loaded with individual voice templates. Consequently, if aircrew voice changes (e.g., under stress) recognition performance is reduced. Moreover some subjects have a much greater natural variability in their voices than others.

d) Continuous speech recognition. Most early requipments can only recognise isolated words, whereas in natural speech the speaker frequently allows one word to flow continuously into the next.

e) Recognition Performance. Even under ideal conditions, recognition scores are always less than 100% and under bad conditions and with poor subjects the scores may be only 50-75%. Thus it is currently necessary and probably will continue to be so-to have some form of feedback to confirm to the speaker that the message has been correctly "captured".

In summary, trials with first generation voice recognition equipments have produced encouraging results, but the need for significant improvements has been identified and these are now being explored. It may be too early to give an exact estimate of the extent to which voice recognition techniques will be used in future combat aircraft, but there is considareble promise that a valuable new interface channel can be developed. First applications are likely to be in areas where 100% accuracy in data transmission is not essential and where an alternative form of data input is also available to aircrew.

3.4 Summary of Requirements

Efficient use of transmission media and restriction imposed on the radio channel capacity by propagational factors and by jamming force the channel bit rate to be kept as low as possible. Under no stress conditions rates ranging from n x 64 kb/s to 2.4 kb/s are required while under heavy jamming, the supportable information rate can go down to 600-200 bit/s. Under all these conditions voice is perceived as a preferred method of man-to-man communications and this requires speech coding from 64 kb/s down to a few hundred bits/s. Sophisticated speech coding methods including variable rate encoding together with Digital Circuit Multiplication (DCM) are invoked also to overcome in the short-to-medium term the areas of economic weakness of 64 kb/s PCM, namely, satellite and long-haul terrestrial links used in switched networks prior to widespread availability of optical fibre links. As far as intra-aircraft communications are concerned machine-to-man (speech synthesis) and man-to-machine (speech recognition) voice communications are considered very necessary because this leaves the hands and the eyes free to perform other functions in the cockpit.

It is to be noted that the military always try to use, to the maximum extent possible, the civil networks which benefit from the economies of scale. There are, however, requirements such as survivability, security, mobility and precedence/pre-emption that are regarded as vital by the military but not considered important for civil applications. The experience shows however that the service features required by the military in time, become requirements also for the civilian systems. This is certainly true as far as the following network features and trends [13] are concerned:

i) Voice coding at sub-rates of 64 kb/s and as low as 16 kb/s and even lower, for long connections and mobile applications.

ii) Speech synthesis and speech recognition using subrates of 64 kb/s, for instance for voice message services and recorded announcements.

iii) Digital Circuit Multiplication (DCM) for making more efficient use of the transmission media.

iv) Long-term objective of integrating voice, data and imagery in the evolving broad-band ISDN [14] when the "Asynchronous Transfer Mode" (ATM) of operation is expected to be implemented using packetised speech. DCM applications are related to the use of digital links at speeds on the order of few Mb/s, while ISDN-ATM applications are foreseen at much higher limit speeds (i.e., 50-150 Mb/s).

There are, however, important differences between the military and civil applications as far as environmental factors are concerned; acoustic noise, vibration, acceleration and jamming are some of them. In the lectures that follow, speech processing will be treated in all its aspects considering both civil and military requirements and applications.

4. SPEECH PROCESSING

4.1 General

Having established the fact that the spoken word plays and will continue to play a significant role in man-man, man-machine and machine-man communications for civil and military applications both real-time and with intermediate storage (e.g., for "voice mail") a brief look will now be taken at the developments in speech proscessing that contribute significantly in all these areas. The problem of speech compression and composition, i.e., Speech Processing arose out of a study of the human voice, for example, Alexander Graham Bell and his father and later Sir Richard Paget [16] and others had studied speech production and operation of the ear [15]. In 1939 Homer Dudley [17,18] demonstrated the Vocoder at the New York World's Fair. This instrument produced artificial voice sounds, controlled by the pressing of keys, and could be made to "speak" when controlled manually by a trained operator. In 1936 Dudley had demonstrated the more important Vocoder; this apparatus gave essentially a means for automatically analysing speech and reconstructing or synthesising it.

The British Post Office also started, at about this date, on an independent program of development, largely due to Halsey and Swaffield [19]. Despite the marginal qualitiy, vocoder was used on High-Frequency radio on, for instance, transatlantic routes, to provide full digital security. The inauguration in 1950 of the first transatlantic undersea cable providing 36 voice circuits (1 million dollars per channel) encouraged work on bandwith conservation. This led to the deployment in 1959 of a speech processing technique known as TASI (Time Assignment Speech Interpolation) which doubled the capacity of the cable by taking advantages of limited voice activity during a call; only the active parts of a conversation (talkspurts) are transmitted.

Efforts in speech processing continued, driven by the requirement to use transmission capacity efficiently, till about 1970 when making computers more useful for humans emerged as the trend spurred by the advances and proliferation of digital computers. This interest centered on the use of human voice for man-computer interaction. Speech synthesis concerns machine-to-man communication (talking machine) and speech rcognition allows machines to listen to and "understand" human speech. Most of the technology for reducing speech bandwith applies to speech synthesis and recognition, but the objective of achieving transmission efficiency still remains as the main motivation for speech processing work despite the promise of very large bandwidth from optical fibres.

Fig 7 shows roughly the relationship between speech transmission and recognition and sythesis of speech [20]. In each case processing starts with "preprocessing" which extracts some important signal characteristics. The following stage which is still a preprocessing stage but extracts more complicated and combinatorial parameters such as segmented phoneme

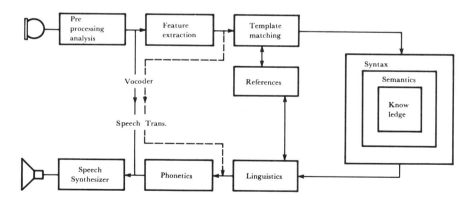

Fig.7. Relation Between Different Speech Bandwidth

Compression and Coding Techniques [20]

parameters or prosodic parameters like speech intonation which are necessary for a speech recognition system. The succeeding stages are corcerned with the central issue of recognition and understanding. A speech output is then produced based on linguistic rules. The phonetic and speech synthesis parts again handle the higher and lower level parameters to produce a speech signal which, when applied to a loudspeaker/earpiece, is converted into an acoustic signal. In a speech transmission system with redundancy reduction (compression), the inner part of Fig 7 is by-passed and a parametic description of the analysed signal is directly sent to a synthesiser which can reproduce the speech signal.

4.2 Speech Coding

Speech compression systems can generally be classified as either Waveform Coders or Vocoders, Fig. 8 (i.e., voice coders or analysis-synthesis telephony). These two classes cover the whole range of compressibility from 64000 down to a few hundred bits per second. The important factors which need to be taken into account when comparing different encoding techniques are the speech quality achievable in the presence of both transmission errors and acoustic noise, the data rate required for transmission, the delay introduced by processing, the physical size of the equipment and the cost of implementation (a function of coder complexity which can be measured by the number of multiply-add operations required to code speech, usually expressed in millions of instructions per second "MIPS").

The most basic type of waveform coding is pulse code modulation (PCM) consisting of sampling (usually at 8 kHz), quantising to a finite number of levels, and binary encoding. The quantiser can have either uniform or non-uniform steps giving rise to linear and logarithmic PCM respectivly. Log-PCM has a much wider dynamic range than linear PCM for a given number of bits per sample, because low amplitude signals are better represented, and as a result logarithmic quantisation is nearly always used in wideband speech communications applications. A data rate of 56 to 64 kbit/s is required for commercial quality speech and lower rates for military tactical quality.

There are many variations on the basic PCM idea, the most common being differential encoding and adaptive quantisation. Each variation has the object of reducing the data rate required for a given speech quality, a saving of approximately 1 bit per sample (8 kbit/s) being achieved when each is optimally employed. In differential PCM (DPCM) the sampled speech signal is compared with a locally decoded version of the previous sample prior to quantisation so that the transmitted signal is the quantised difference between samples. In adaptive PCM (APCM)the quantiser gain is adjusted to the prevailing signal amplitude, either on a short term basis or syllabilcally. By controlling the adaption logic from

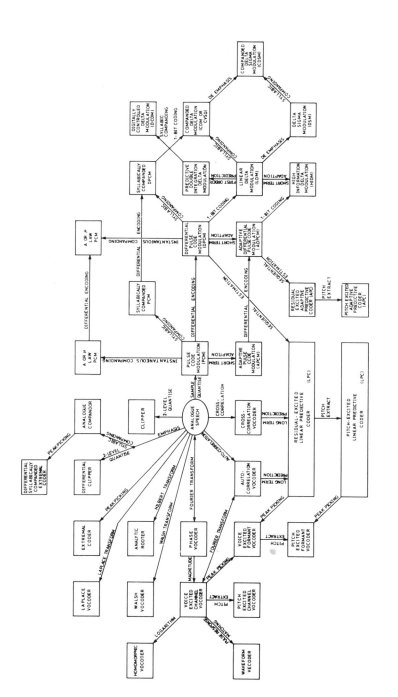

Fig.8. Relationship Between Different Speech Bandwith Compression and Coding Techniques

the quantiser output, the quantiser gain can be recovered at the receiver without the need for additional information to be transmitted. Adaptive differential PCM (ADPCM) is a combination of DPCM and APCM which saves 2 to 4 bits per sample compared with PCM, thus giving 48 to 32 kb/s with high quality speech.

It is interesting to note that although the principle of DPCM has been known for 30 years, it was not possible to standardise such a 32 kb/s coder until 1983 [21], after efficient and robust algorithms became available. These adaptive algorithms are efficient in the sense that they adapt quantisation and prediction synchronously at the encoder and decoder without transmitting explicit adaptation information. They are robust in the sense that they function reasonably well even in moderate bit-error environment.

There is another adaptive approach to producing high quality and lower bit-rate coder which is called "adaptive subband coding" which divides the speech band into four or more contiguous bands by a bank of filters and codes each band using APCM. After lowering the sampling rates in each band, an overall bit rate can be obtained while maintaining speech quality; by reducing the bits/sample in less perceptually important high-frequency bands. Bands with low energy use small step sizes, procuding less quantisation noise than with less flexible systems. Furthermore, noise from one band does not affect other frequency bands. Coders operating at 16 kb/s using this technique have been shown to give high quality but with high complexity [22].

When the number of quantisation levels in DPCM is reduced to two, delta modulation (DM) results. The sampling frequency in this case is equal to the data rate, but it has to be well above the Nyquist frequency to ensure that the binary quantisation of the difference signal does not produce excessive quantisation noise. Just as with PCM, there are many variations of DM, and the right hand side of Fig 8 illustrates some of them. The most important form of DM used in digital speech communications is syllabically companded DM; there are a number of closely related versions of this, examples being continuously variable slope DM (CVSD) and digitally controlled DM (DCDM). The data rate requirements are a minimum of about 16 kb/s for military tactical quality speech and about 48 kb/s for commercial quality.

When operated at data rates of 12 kbit/s and lower, the speech quality obtained with PCM and DM coders is poor, and consequently they cannot be used as narrow band devices. However, the principles of operation of wideband coders are useful in analysis-synthesis telepnony once significant redundancy has been removed from the speech waveform. Examples of this are digital encoding for the transmission of individual speech parameters and the relationship between LPC and DPCM indicated in Fig 8.

Analysis-synthesis telephony techniques are based on a model of speech production. Fig 9 (a) shows a lateral cross-section through the human head, and illustrates the various organs of speech production. Briefly, these are the vocal tract running from the vocal chords at the top of the larynx to the mouth opening at the lips, and the nasal tract branching off the vocal tract at the velum and running to the nose opening at the nostrils. The glottis (the space between the vocal chords) and the sub-glottal air pressure from the lungs together regulate the flow of air into the vocal tract, and the velum regulates the degree of coupling between the vocal and nasal tracts (i.e., the nasalisation).

There are two basic types of speech sound which can be produced, namely voiced and unvoiced sounds. Voiced sounds occur when the vocal chords are tightened in scuh a way that the subglottal air pressure forces them to open and close quasi-periodically, thereby generating "puffs" of air which acoustically excite the vocal cavities. The pitch of voiced sounds is simply the frequency at which the vocal chords vibrate. On the other hand, unvoiced sounds are produced by forced air turbulence at a point of constriction in the vocal tract, giving rise to a noise-like excitation, or "hiss".

A model of speech production often used for the design of analysis-synthesis vocoder is shown in Fig. 9 (b). In this model, a number of simplifications have been made, the most important ones being that the excitation source for both voiced and unvoiced sounds is located at the glottis, that the excitation waveform is not affected by the shape of the vocal tract, and that the nasal tract can be incorporated by suitably modifying the vocal tract. These simplifications lead to differing subjective effects, depending on the type of speech sound and the particular vocoder being used.

In channel vocoding the speech is analysed by processing through a bank of parallel band-pass filters, and the speech amplitude in each frequency band is digitized using PCM techniques. For synthesis, the vocal and nasal tracts are represented by a set of controlled gain, lossy resonators, and either pulses or white noise are used to excite them. In pitch-excited vocoders, the excitation is explicitly derived in the analysis, whereas in voice-excited vocoders it is derived by non-linear processing of the speech signal in a few of the low frequency channels combined into one. Pitch-excited vocoders require data rates in the range from 1200 to 2400 bit/s and yield poor quality speech, whereas voice-excited vocoders will provide reasonable speech quality at 4800 bit/s and good quality at 9600 bit/s.

A formant vocoder is similar to a channel vocoder, but has the fixed filters replaced by formant tracking filters. The centre frequecies of these filters along with the corresponding speech formant amplitudes are the transmitted parameters. The main problem is in acquiring and maintaining lock on the relevant spectral peaks during vowel-consonant-vowel transitions, and also during periods where the formants become ill-defined. The data rate required for formant

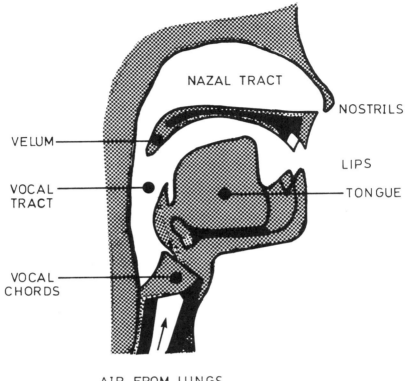

Fig.9(a). Lateral cross-section of human head

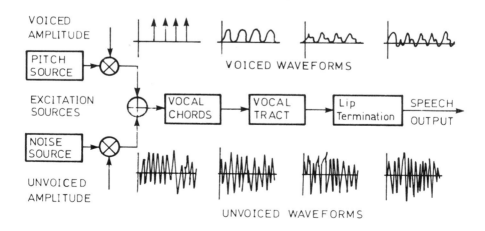

Fig.9(b). Simplified Model of Speech Production

vocoders can be as low as 600 bit/s, but the speech quality is poor. The minimum data rate required to achive good quality speech is about 1200 bit/s, but to date this result has only been obtained using semi-automated analysis with manually interpolated and corrected formant tracks.

The third method of analysis-synthesis telephony to have achieved importance is linear predictive coding(LPC). In this technique the parameters of a linearised speech production model are estimated using mean-square error minimisation procedures. The parameters estimated are not acoustic ones as in channel and formant vocoders, but articulatory ones related to the shape of the vocal tract. For a given speech quality, a transmission data rate reduction in comparison with acoustic parameter vocoding should be achieved because of the lower redundancy present. Just as wtih channel and formant vocoders, excitation for the synthesizer has to be derived from a separate analysis, the usual terminology being pitch-excited or residual excited, corresponding to pitch or voice excitation in a channel vocoder. LPC is a very active area of speech research, and new results appear regularly. At persent data rates as low as 2400 bit/s have been achieved for pitch-excited LPC with reasonable quality speech, and in the range from 8 kbit/s to 16 kbit/s for residual excited LPC with good speech quality.

The application of vector quantisation (VQ), a fairly new direction in source coding, has allowed LPC rates to be dramatically reduced to 800 b/s with very slight reduction in quality, and further compressed to rates as low as 150 b/s while retaining intelligibility [23,24]. This technique consists of coding each set or vector of the LPC parameters as group instead of individually as in scalar quantisation. Vector quantisation can be used also for waveform coding.

A good candidate for coding at 8 kb/s is multipulse linear predictive coding, in which a suitable number of pulses are supplied as the excitation sequence for a speech segment-perhaps 10 pulses for a 10-ms segment. The amplitudes and locations of the pulses are optimised, pulse by pulse, in a closed-loop search. The bit rate reserved for the excitation information is more than half the total bit rate of 8 kb/s. This does not leave much for the linear predictive filtre information, but with VQ the coding of the predictive parameters can be made accurate enough.

For 4 kb/s coding, code excited or stochastically excited linear predictive coding is promising. The coder stores a repertory of candidate excitations, each a stochastic, or random sequence of pulses. The best sequence is selected by a closedloop search. Vector quantization in the linear predictive filter is almost a necessity here to guarantee that enough bits are available for the excitation and prediction parameters. Vector quantization ensures good quality by allowing enough candidates in the excitation and filter codebooks.

Table II below compares tradeoffs for representative types of speech coding algorithms [25]. It shows the best overall match between complexity, bit rate and quality. A coder type is not necessarily limited to the bit rate stated. For example, the medium-complexity adaptive differential pulse-code modulation coder can be redesigned to give communication-quality speech at 16 kb/s instead of high-quality speech at 32 kb/s. In fact, a highly complex version can provide high-quality speech at the lower bit rate. Similarly lower-complexity multipulse linear predictive coding can yield high-quality coding at 16 kb/s, and a lower-complexity stochastically excited linear predictive coder (LPC) can be designed if the bit rate can be 8 kb/s instead of 4 kb/s.

Table II: Comparison Low Bit-Rate Speech Coding Schemes

Coder type	Bit rate kb/s	Complexity MIPS	Delay ms	Quality	MOS
*Pulse-code modulation	64	0.01	0	High	
*Adaptive differential pulse-code modulation	32	0.1	0	High	> 4
*Adaptive subband coding	16	1	25	High	
*Multipulse linear predictive coding	8	10	35	Communication	
*Stoshastically excited linear predictive coding	4	100	35	Communication	> 2
*LPC vocoder	2	1	35	Syntetic	< 2

Cost is also a tradeoff factor, but it is hard to quantify in a table. The cost of coding hardware generally increases with complexity. However, advances in signal processor technology tend to decrease cost for a given level of complexity and, more significantly, to reduce the cost difference between low-complexity and high-complexity techniques.

Of course, as encoding and decoding algorithms become more complex they take longer to perform. Complex algorithms introduce delays between the time the speaker utters a sound and the time a coded version of it enters the transmission systems. These coding delays can be objectionable in two-way telephone conversations, especilaly when they are added to delays in the transmission network and combined with uncanceled echoes. Coding delay is not a problem if the coder is used in only one stage of coding and decoding, such as in voice storage. If the delay is objectionable because of uncanceled echoes the addition of an echo canceler to the voice coder can eliminate or mitigate the problems. Finally, coding delay is not a concern if the speech is merely stored in digital form for later delivery.

Many explanations can be given as to why particular types of speech coder do not perform well at low data rates. With waveform coders, it is generally accepted that the main reason is excessive quantisation noise despite companding and/or adaptive logic. With analysis-synthesis techniques, the main reasons are over-simplification of the vocal tract model, leading to imprecise spectral characterization, and unreliable pitch detection and voiced-unvoiced-silence decisions in the analyser which, coupled with an over-simplified excitation model in the synthesizer, lead to imprecise temporal characterisation and a lack of naturalness in the synthetic speech.

In conclusion on speech coding, it should be remarked that there are two complementary trends that are at work in digital telecommunications: speech coding developers are trying to reduce the bit rate for a given quality level while developers of modulation and demodulation tehcniques are endeavoring to encrease the bit rate that a channel of a given bandwith can accomodate.

The limiting capacity C (b/s) of a channel with a bandwith B and the signal-to-noise ratio (SNR) is given by Shannon's theory of communication as

$$C = B \cdot \text{Log}_2 (1 + \text{SNR})$$

A typical analogue telephone channel with B = 3 kHz and SNR = 30 dB would therefore have C = 30 kb/s. A modulation system with this performance has yet to be devised however.

Limiting performance for speech coding may be calculated as follows. In the English language there are $42 = 2^{5.4}$ distinct sounds called "phonemes" [26], and normal speech is basically a continues process of interpolation between these sounds. A normal talker utters about ten phonemes per second [27], and the basic information of speech (the information rate of the written equivalent of the words spoken) is thus only about $5.4 \times 10 = 54$ b/s. If one allows for, say, 1560 variations on the basic phonemes to accommodate different dialects and personal characteristics, then the total number of sounds is $42 \times 1560 = 2^{16}$, and if one allows for a very fast talker uttering, say 40 phonemes per second, then the information rate is still only about $40 \times 16 = 640$ b/s. There is thus a large discrepancy between the data rate required for a good quality PCM system and the rate at which real information is transmitted. The stage of development at present is such that it may soon be possible to send high-quality digital speech signals at about 8 kb/s over a wide range of channels. Robust, high-quality coding algorithms will cut the bit rate and new modulators and demodulators will transmit the lower bit-rate, with a low bit-error probability over an analogue channel having a bandwith of about 3 kHz. Analogue voice link, now used for transmitting high-quality analogue speech, will therefore be able to carry high-quality digital speech with added benefits as voice security.

This rather lengthy precis on speech coding which is given here because of the central role of the subject in the whole speech processing field will be elaborated on and expanded by Prof. Gersho in Chapter 3 on "Speech Coding".

4.3 Speech Synthesis

Speech synthesis involves the conversion of a command sequence or input text (words or sentences) into speech waveform using algorithms and previously coded speech data. The text can be entered by keyboard, optical character recognition, or from a previously stored data base. Speech synthesizer can be characterized by the size of the speech units they concatenate to yield the output speech as well as by the method used to code, store and synthesize the speech. Large speech units, such as phrases and sentences can give high-quality output speech (with large memory requirements). Efficient coding methods reduce memory needs, but usually degrade speech quality.

Synthesizers can be divided into two classes: text-to-speech systems which constructively synthesize speech from text using small speech units and extensive linguistic processing, and voice response systems which reproduce speech directly from previously-coded speech, primarily using signal processing techniques. Voice response systems are often called "speech coders" and contain both an analyzer and a synthesizer.

Synthesizers can also be classified by how they parametrize speech for storage and synthesis. High quality systems with large memory capacities synthesize speech by recreating the waveform sample-by-sample in the time domain. More efficient (but lower quality) systems attempt to recreate the frequency spectrum of the original speech from a parametric representation. A third possiblitiy is direct simulation of the vocal tract movements using data derived from X-ray analysis of human production of specified sound sequences.

Due to the difficulty of obtaining accurate three dimensional vocal tract representations modeling the system with a limited set of parameters, this last method usually yields lower quality speech and has yet to have commercial application.

The simplest synthesizers concatenate stored words or phrases. This method yields high-quality speech (depending on the synthesis method) but is limited by the need to store in computer (read-only) memory all the phrases to be synthesized after they have been spoken either in isolation or in carrier sentences. For maximum naturalness in the synthetic speech, each word or phrase must originally be pronounced with timing and intonation appropriate for all sentences in which it could be used.

Hybrid synthesisers concatenate intermediate-sized units of stored speech such as syllables, demisyllables, and diphones, using smoothing of special parameters at the boundaries between units. To further enhance the flexibility of stored-speech synthesis systems, one can allow control of prosody (pitch and duration adjustments) during the synthesis process. With the decreasing cost of digital storage, stored-speech synthesis techniques could provide low-cost voice output for many applications.

It is clear that stored-speech systems are not flexible enough to convert unrestricted English (or whatever language) text to speech. A text-to-speech system that uses synthesis-by rule is needed for applications such as accessing electronic mail by voice, a reading machine etc. The text-to-speech system must convert incoming text, such as electronic mail, that often includes abbreviations, Roman numerals, dates, times, formulas, and a wide variety of punctuation marks into some reasonable, standard form. The text must be further translated into a broad phonetic transcription. How this is done and other aspects of speech synthesis are explained in Chapter 4 by Dr Flanagan.

There are several commercial text-to-speech conversion systems in the market which come in board, peripheral, software or system form [28]. They are mostly for English adult male but some do adult female and child voice. The speech mode used is mostly words with some accepting also letters. The synthesis technique employed is mostly formant synthesis but some manufactures use LPC. Prices vary from a few hundred Dollars for software to a few tens of thousand Dollars for systems. The quality of even the best systems is such that during tests, listeners understood the synthetic speech produced 97.7% of the time compared with 99.4% for human speech. Research in text-to speech synthesis which concentrates, at present, on producing speech that sounds more natural, is expected to provide systems which are more flexible for selecting the speaker characteristics, different languages and their dialects, and regional variabilities.

4.4 Speech Recognition

Of all the speech processing techniques, speech recognition is the most intractable one. The ultimate objective of most research in this area is to produce a machine which would understand conversational speech with unrestricted vocabulary, from essentially any talker. We are far from this goal.

The reason why automatic speech recognition is such a difficult problem can be stated very briefly under four problem areas: First, the speech signal is normally continuous and there are no acustic markers which identify the word boundaries. Second, speech signals are highly variable from person to person and even in one and the same person depending on his state. The third problem area is ambiguity which is characterised by conditions whereby patterns which should be differenet end up looking alike. The fourth problem area results from the fact that the speech signal is a part of the complex system of human language where it is often the intention behind a message that is more important than the message itself. Therefore an advanced speech recogniser would be expected to incorporate techniques which would enable it to use the meanings of words in order to interpret what has been said. However, there are several applications which do not require this full capability. They range from voice editors, and information retrieval from data bases to basic English and large vocabulary systems required for office dictation/word processing and language translation.

A technology that is closely related to speech recognition is speaker recognition, or automatic recognition of a talker from measurements of individual characteristics in the voice signal. The two tasks that are relevant here are "absolute identification" and "talker verification" the former being the more difficult to perform. An interesting military application of speaker recognition is related to the monitoring of enemy radio channels with a view to identifying, perhaps in conjunction with keyword recognition, critical situations before they occur.

The recognition problem has at least three dimensions: vocabulary size, speaker identify and fluency of input speech and the performance of speech recognisers also depend on the acoustic environment and transmission conditions. Current understanding permits building practical systems that reliably recognise several hundred words spoken by a person who trained the system. Recognition for any or all speakers requires about ten times more computation than for individuals whose vocabulary patterns have been stored. Recognition of single words or short phrases-spoken in isolation-can be done reliably, even over dialed-up telephone channels. Recognition of connected words is under active development. Recognition of conversational fluent speech is in fundamental research, and advances strongly depend on good computational models for syntax and semantics.

Dr Rabiner reviews and discusses in Chapter 5 the general pattern recognition framework for machine recognition of speech including some of the signal processing and statistical pattern recognition aspect. He comments on the performance of current systems and also on the way ahead in this very challenging area. He shows that our understanding is best for the simplest recognition task and is considerably less well developed for large scale recognition systems.

5. QUALITY EVALUATION METHODS

There are different and as yet genarally not standardised methods (subjective and objective) to measure the "goodness" or "quality" of speech processing systems in a formal manner. The methods are divided into three groups:

- Subjective and objective assessment for speech coding and transmission systems.
- Subjective and objective quality measures for speech output systems (synthesisers).
- Assessment methods for automatic speech recognition systems.

Mr Steeneken discusses in Chapter 6 assessment methods for these three groups of speech processing systems. The first two systems require an evaluation in terms of intelligibility measures while the evaluation of speech recognisers requires a different approach as the recognition rate normally depends on recogniser-specific parameters and external factors. However, more generally applicable evaluation methods such as predictive methods are also becoming available. For military applications it is, of course, necessary to include into the test method the effects of the environmental conditions such as noise level, acceleration, stress, mask microphones etc. It is emphasised that evaluation techniques are crucial to the satisfactory deployment of speech processing equipments in real applications.

Because of its widespread use and to define some important speech parameters, a subjective assesment method, which is generally used to measure the perceived speech quality of a speech coder, is outlined below.

The term "quality" is a general term combining many different attributes, and there are many ways in which these can be assesed. The most important attributes contributing to speech quality are:

- intelligibility (a measure of "understandableness")
- articulation score (a measure of phoneme recognition)
- speaker identification

5.1. Intelligibility

The most well known technique for measuring intelligibility is the Harvard Test [29]. This test consists of transmitting list of phonetically balanced (PB) words through the speech coder under test, and measuring the proportion of words correctly perceived. The PB word lists consist of isolated but meaninful words; they are selected in such a way that each phoneme contained in the list has the same probability of occurrence as it has in normal conversational speech.

An alternative method of measuring intelligibility is to use meaningful sentences rather than PB word list. The percentage of words correctly perceived then gives a measure of intelligibility. Note that the intelligibility when sentences are used is higher than that obtained by using PB word lists because the meaning associated with sentences gives perceptual clues to the listener and these clues are not available with PB word lists. When reporting intelligibility scores it is therefore important to specify which type of test was used and under what conditions it was conducted.

5.2 Articulation score

The intelligibility tests outlined in the previous section measure the degree of speech understanding available with a particular speech coder. If the intelligibility is high, however, (e.g. more than 90%) then the tests are not very sensitive to small differences between different types of coder. A more sensitive test is to measure the articulation score instead of the intelligibility. The increase in sensitivity could be achieved by using logatoms (i.e., nonsense syllables) instead of words [30]. The chosen logatoms could be phonetically balanced for all phonemes or for the consonants only. The articulation score derived from a consonant recognition test (CRT) using the latter type of logatom would perhaps give the most meaningful intelligibility measure for military applications, because the main clues in perception are derived from consonants rather than vowels. An illustration of this is the sentence

-a- -ou u--e--a-- --i-

in which all the consonants have been replaced by a hyphen. It is not very meaningful. If the opposite condition, in which all the vowels instead of the consonants are replaced by a hyphen, is now applied to the same sentence one has

c-n y-- -nd-rst-nd th-s

which is much more meaningful. The full sentence is of course:

can you understand this

5.3 Speaker Identification

The ability of a speech coder to transmit the characteristics of a speaker's voice in such a way that a listener can identify who is speaking is another attribute contributing to perceived speech quality. In a military environment, this is an important attribute because of the "need-to-know" principle.

There is basically only one method for measuring the speaker identification capability of a speech coder and that is simply to use a number of different speakers and intruct the listeners to identify which speaker they think they are hearing. The percentage of correct estimates then yields a measure of the speaker identification ability of the particular speech coder under test.

5.4 Quality

The combined effects of the attributes outlined in the previous three sections (i.e., intelligibility, articulation score, and speaker identification) can best be measured by conducting "user opinion tests" [31]. Such tests simply consist of instructing a pair of users to discuss a given problem for a certain period of time via the speech coder under test, and then to ask them to classify their opinion in terms of a five-point scale given in Table III below. The results obtained from user opinion tests, averaged over a large number of users, yield an indication of the overall speech quality of the speech coder under test.

Table III: Five-Point Adjectival Scale for Quality Impairment and Associated Number Scores

Number Scores	Quality	Impairment Scale
5	Excellent	Imperceptible
4	Good	Perceptable but not annoying
3	Fair	Slightly annoying
2	Poor	Annoying
1	Unsatisfactory	Very annoying

5.5 Measurement

An alternative method for quantifying the "goodness" of a speech coder, other than assessing the rather ill-defined concept of speech quality is to measure its electrical characteristics. Important characteristics which could be measured include:
- attenuation frequency distortion
- signal-to-noise ratio
- dynamic range
- idle channel noise
- quantization noise
- susceptibility to transmission errors
- harmonic distortion
- group delay distortion

In order to combine the results of such measurements into a single entity indicating the "goodness" of the coder, an "articulation index" (AI) could be computed [32]. If so desired, this index might then be directly related to either an articulation score or an intelligibility assesment. The validity of such a relationship and the method used for calculating the AI, are still topics of research and development but very promising results have already been achieved [33].

6. THE SPEECH SIGNAL

In discussing Speech Processing techniques one must, of course, be fully aware of and take into account how humans generate the speech signal, how they perceive it and the process of speech communication itself.

Dr Hunt deals in Chapter 2 with these subjects which underpin all the other chapters and shows the problem areas with which the researchers in the speech processing area are faced. He presents speech communication as an interactive process, in which the listener actively reconstructs the message from a combination of acoustic cues and prior knowledge, and the speaker takes the listener's capacities into account in deciding how much acoustic information to provide.

7. CONCLUSIONS

Speech communication is and will remain in the foreseeable future the main mode of communication, not only for civil but also for strategic/tactical military applications. Digital speech processing is, consequently, an essential ingredient of the evolving ISDN's to be used by both civil and military users. End-to-end digital connections of the kind promised by ISDN are well suited to secure communications. Furthermore, the ubiquity, connectivity, and interoperability inherent in the concept will be most valuable in emergency situations requiring reconfigured communications.

Speech coding methods have been standardised internationally at 64 kb/s (PCM) and 32 kb/s (ADPCM) and coders at these rates are being used in the common-user switched telephone networks. Continuously Variable Slope Delta Modulation (CVSD) has also been standardised in NATO for tactical military communications. There are also both civil and military requirements for speech coders operating at speeds of 16 kb/s and below e.g., for mobile land and maritime communications. For HF communications and LOS radio and satellite communications under heavy jamming, vocoders operating at 2.4 kb/s and even below are required. Secure voice using 4 kHz nominal analogue channels also requires speech coders operating at speeds of 4.8 kb/s and below. Speech coding is also required for high-fidelity voice (HFV) with 7 and 15 kHz bandwith as well as for Digital Circuit Multiplication and for longer-term applications, i.e., in the evolving broadband ISDN when "Asynchronous Transfer Mode" (ATM) of operation will be implemented. All these subjects are dealt with in Chapter 7.

It is to be noted that there are important operational requirements for interoperability between systems using different speech coders; this necessitates standardisation and agreements on interfaces/gateways where code, rate and other (signalling, numbering) conversions take place.

To achieve good quality below 32 kb/s codes must take increasing advantage of the constraints of speech production and perception. At transmission rates below 16 kb/s quality diminishes significantly, requiring more of the, as yet, poorly known properties of speech production and perception. Also at the lower transmission rates, the computational complexity to implement the coding algorithms increases, while the ability to handle nonspeech-like sounds-such as music and voice-band data diminishes. Typically too, the encoding delay increases as the transmission bit rate decreases.

The primary challenge, then is to develop new understanding that will significantly elevate the speech-quality curve for the lower bit rates, even with substantial but acceptable increase in complexity.

The research frontier in coding currently centers on ways to achieve good quality at transmission rates of 9.6 kb/s and below. Undoubtedly, increased computational complexity will be required to elevate the quality of low bit-rate codes, which must extensively use the known redundancies of speech production and perception. Breakthroughs will occur only when new properties of redundancy are found [34].

In addition to speech coding, there are evolving Command and Control requirements for speech synthesis and speech recognition systems on the ground as well as in the cockpit involving voice storage, voice response, voice control, speaker authentication/recognition etc. These systems are expected to find important applications also in the civil networks [34]: telephone answering, remote access, voice mail, speaker verification etc.

In speech synthesis, first systems for unrestricted text-to-speech conversion are producing useful, intelligible synthetic speech but of limited naturalness. Over the next five years, work already in progress aims to produce high-quality synthesis from text, where different voice qualities (such as man, woman, child) might be specified. Also, synthesis from text might be realized for languages that are quite different from Western languages. Over the long term, detailed understanding may permit specifying individual voice characteristics, dialects, and accents.

In speech recognition, systems for reliable recognition of isolated words are well established and beginning to prove their value. The near term will see speaker-independent recognition of connected digits established and applied.

Over the next few years, the technology is expected to advance to whole connected sentences, using limited vocabularies and finite grammars. Over the longer term, understanding of programmed parsers and natural language analysis will allow the leverage of syntax, semantics and eventually, even pragmatics to expand a machine's conversational ability. Ultimately, practical spoken language translation may be possible.

While research and development work, driven by the advances being made in the areas of microelectronics, computer science, and artificial intelligence, continue vigorously in many national laboratiories on all aspects of speech processing, international/regional/national standardisation bodies try to promulgate standards in order to achieve the necessary or desired degree of uniformity in design or operation to permit voice systems to function beneficially for both providers and users.

NATO, as a body is involved in standardisation efforts through its "Military Agency for Standardisation" (MAS) which has already issued "Standardisation Agreements" STANAG's on 2.4, 4.8 and 16 kb/s coders and modulation equipment. Also within NATO, the member countries are engaged in active technical coordination, information exchange and cooperative research projects through the NATO AC/243 Panel III Research Group (RSG)-10 for speech processing. The activites of this Group include, among other things, the application of speech input/output systems in the multilingual military environment. The countries that participate in the work of RSG-10 are Canada, France, Germany, Netherlands, United Kingdom, and the United States. In fact, two of our authors are members of this Group.

This book is a state-of-the art review of speech processing written by scientists who are in the forefront of research in this facinating area, and the Editor of this book would feel gratified if this results in inducing or seducing some of the readers into this area of work or in fertilising their own fields of expertise.

8. REFERENCES

[1] Cherry C., "On Human Communication", Science Editions, Inc., New York, 1961

[2] Ince A.N., et al., "The System Master Plan and Architectural Design Studies for TAFICS", PTT-TAPO Technical Reports, Ankara, 1989.

[3] Ochsman R.B. et al, "The Effects of 10 Communications Modes on the Behavior of Teams During Cooperative Problem Solving", International Journal Man-Machine Studies, Vol 6, 1974.

[4] Smith M., "A Model of Human Communication", IEEE Com. Magazine, Vol 26, No 2, Feb. 1988.

[5] Drukarch C.Z. et al., "X.25: the Universal Packet Network", Proc. Fifth Int. Conf. Comput. Commun.", Oct. 1980

[6] Weir D.F., Holnblad J.B., Rothberg A.C., "An X.75 Based Network Architecture", Proc. Fifth Int. Conf. Comput. Commun." Oct.1980.

[7] Ross M. and Mowafi D., "Performance Analysis of Hybrid Switching Concepts for Integrated Voice/Data Communications", IEEE Trans. Com mun., COM-30 No.5, May 1982.

[8] Haselton E.F., "A Per Frame Switching Concept Leading to Burst Switching Network Architecture", Proc. ICC 1983.

[9] Gitman I. and Frank H., "Economic Analysis of Integrated Voice and Data Networks; A Case Study", Proc. IEEE Vol.65, Nov.1978.

[10] Heggestad H.M. and Weistein C.J., "Voice and Data Communications Experiments on a Wideband Satellite/Terrestrial Interwork System", Proc. Int. Conf. Commun., Boston, MA, June 1983.

[11] Falk G. et al., "A Mulutiprocessor Channel Schedular for the Wideband Packet Satellite Network" Proc. Int.Conf. Commun., Boston, MA, June 1983.

[12] "The potential Impact of Developments in Electronic Technology on the Future Conduct of Air Warfare", AGARD Advisory Report No.232, Vol 3,1986.

[13] "CCITT SG XVIII, Rep. R.17, Working Party 8", Geneva Meeting, March 1986.

[14] "CCITT SG XVIII, Rep. R.45, Working Party 8", Hamburg Meeting July 1987.

[15] "Encyclopedia Britanica", Cambridge University Press, London, 11th Ed. 1911.

[16] Paget R., "Human Speech", Kegan Paul, Trench, Trubner and Co.Ltd. London 1930.

[17] Dudley H., "The Carrier Nature of Speech", Bell System Tech.J., 19 Oct.1940.

[18] Dudley H., et al, "A Synthetic Speaker", J.Franklin Inst., 227, 1939.

[19] Halsey R.J. and Swafield J., "Analysis-Synthesis Telephony with Special Reference to the Vocoder", J.Inst. Elec. Engrs (London), 95, Part III, 1948.

[20] Mangold H., "Analysis, Synthesis and Transmission of Speech Signals", AGARD Lecture Series No.129 "Speech Processing", May 1983.

[21] "CCITT Red Book, Vol III", Geneva, ITU Press, 1948.

[22] Crochiere R.E., et al., "Real-Time Speech Coding", IEEE Trans. On Commun., Vol COM-30, April 1982.

[23] Buze A., et al., "Speech Coding Based upon Vector Quantisation"., IEEE Trans. Acoust., Speech and Signal Process., ASSP-28, No.5, Oct.1980.

[24] Roucos S., et al., "Vector Quantization for Very-Low-Rate Coding of Speech", Conf.Rec., 1982 IEEE Global Coms Conf., FL, Nov.29-Dec.2, 1982.

[25] Jayant N.S., "Coding Speech at Low Bit Rates", IEEE Spectrum, Aug. 1986.

[26] "Principles of the International Phonetic Association", Dept. of Phonetics, University Collage, London, 1949.

[27] Flanagan J.L., "Speech Analysis Synthesis and Perception", Springer Verlag, Berlin, 1972.

[28] Kaplan G. and Lerner E.J., "Realism in Synthetic Speech", IEEE Spectrum, April 1985.

[29] "USA Standard Method for Measurement of Monosyllabic Word Intelligibility", American National Standards Institute Inc., New York, 1960.

[30] Fairbanks G., "Voice and Articulation Drill Book", Harper and Bros., New York, 1940.

[31] Ochiai I., "Phoneme and Voice Identification Using Japanese Vowels", Language and Speech, Vol.2, 1959.

[32] Beranek L.L., "Acoustics", McGraw Hill, 1954.

[33] French N.R., Steinberg J.C., "Factors Governing the Intelligibility of Speech Sounds", J.Acoust. Society of America, Vol.19, 1947.

[34] "Speech Processing Technology", ATT Technical Journal, Sept.Oct. 1986 Vol 65, Issue 5.

[35] Ince A.N. and Schemel R. "Factors Affecting Use and Design of Spread-Spectrum Modems for the HF Band", Proc. IEE, Vol. 133, Part F, No. 2, April 1986.

[36] Ince A.N. et al., "Considerations for NATO Satellite Communications in the Post 2000 Era", AGARDograph 330, 1991.

ANNEX

CALCULATION OF SATCOM LINK CAPACITY UNDER JAMMING

The total uplink data rate R_{du} that can be supported by a transmitting SATCOM terminal in the presence of uplink jamming, while maintaining a minimum acceptable uplink E_b/N_o is given by,

$$R_{dU} = \frac{1}{M_U \cdot (E_b/N_o)_U} \cdot \frac{P_T}{\dfrac{kT_S L_U}{G_{RS}} + \dfrac{P_{JU}}{\alpha B_{SU}}}$$

P_T = SATCOM terminal EIRP

P_{JU} = uplink jammer EIRP

G_{RS} = satellite receive antenna gain in the SATCOM terminal direction

α = satellite receive antenna nulling in the jammer direction

T_S = effective noise temperature of satellite receiver

k = Boltzmans constant

B_{SU} = uplink spreading (hopping) bandwidth

L_U = uplink free space loss

$(E_b/N_o)_U$ = minimum acceptable energy per bit-to-noise density ratio after dehopping at the satellite

M_U = margin for atmospheric and rain losses at uplink frequency

In the equation above, the satellite range from the terminal and from the jammer (and hence the uplink free space losses) have been assumed to be equal.

CHAPTER 2

THE SPEECH SIGNAL

Melvyn J. Hunt

Marconi Speech & Information Systems
Airspeed Road, The Airport
Portsmouth, Hants
PO3 5RE
England

ABSTRACT

This chapter provides a non-mathematical introduction to the speech signal. The production of speech is first described, including a survey of the categories into which speech sounds are grouped. This is followed by an account of some properties of human perception of sounds in general and of speech in particular. Speech is then compared with other signals. It is argued that it is more complex than artificial message bearing signals, and that unlike such signals speech contains no easily identified context-independent units that can be used in bottom-up decoding. Words and phonemes are examined, and phonemes are shown to have no simple manifestation in the acoustic signal. Speech communication is presented as an interactive process, in which the listener actively reconstructs the message from a combination of acoustic cues and prior knowledge, and the speaker takes the listener's capacities into account in deciding how much acoustic information to provide. The final section compares speech and text, arguing that our cultural emphasis on written communication causes us to project properties of text onto speech and that there are large differences between the styles of language appropriate for the two modes of communication. These differences are often ignored, with unfortunate results.

INTRODUCTION

This chapter deals with the nature of the speech signal: the signal that allows one human being to communicate to another whatever message he or she consciously chooses to express, with no external aids and usually with very little effort. One of its principal aims is to argue that speech is an exceedingly special kind of signal.

A newly invented or newly discovered signal can be approached objectively. But we can all speak, and the internal impression we have of speech can cloud our view. To compound the problem, most of us can read, and the impression that we gain of language from printed text often distorts our ideas of spoken language. The extent of these problems will be discussed in sections 4 and 5; but before that some more basic information on the production and perception of speech needs to be presented. Properties of both production and perception are exploited in almost all systems for recognition, synthesis and efficient transmission of speech.

THE PRODUCTION OF SPEECH

It may not be obvious why the recognition, artificial generation, and efficient transmission of the speech signal should be helped by an understanding of humans produce it. We do not, after all, need to know how a teleprinter signal was generated in order to transmit it, decode it or reproduce it. Nevertheless, arguments for looking closely at human speech production will emerge towards the end of this section and in later sections. For the moment, we can at least note that production mechanisms provide a useful framework for describing the speech signal.

The following brief account of speech production is simplified in two ways. First, it omits certain production mechanisms not generally found in major European languages, where the airflow is inwards rather than outwards, or where the airflow is driven by some process other than expansion or compression of the lungs: the clicks used in some southern African languages are an example of both of these unusual features. Second, it presents a classical view of distinctions occurring in carefully produced speech. Later sections will present examples where real speech differs from the simple description.

The organs primarily involved in producing speech are the *larynx*, visible as the "Adam's apple" in men, and which contains a pair of muscular folds called the vocal cords, and the *vocal tract*, which is a tube leading from the larynx along the pharynx and then branching into the oral cavity leading to the lips and through the nasal cavity to the nostrils. The nasal side branch can be closed off by raising a valve at the back of the mouth called the *uvula*.

Acoustic energy in speech can be generated in two different ways. The primary mechanism, known as *voiced* excitation, occurs in the larynx. The muscles in the larynx place the vocal cords close together and make them loose enough that when air from the lungs is driven through them they open and close quasi-periodically at an average rate of about 110 times a second for a man and about twice that for a woman. The main instant of voiced excitation occurs not, as one might expect, on opening, but when the airflow from the lungs is suddenly stopped as the cords are pulled together by Bernouilli forces. The resulting voiced speech sounds include all vowels (unless whispered) and many consonant sounds: the words *Roman, yellow,* and *wiring,* for example, are composed entirely of voiced sounds.

In the second mechanism for generating acoustic energy in speech air passes from the lungs through the larynx with the vocal cords held apart and is forced through a constriction formed by the tongue or lips causing turbulence and resulting in an aperiodic, noise-like excitation. Sounds generated purely in this way (such as the "s" and "ft" in *soft*) are said to be *voiceless,* and they generally play a less important role in speech than voiced sounds.

The two excitation mechanisms just described can occur simultaneously, as they do in the initial sounds of *zip* and *vat.* In English, at least, sounds with both kinds of excitation constitute the smallest of the three classes.

As we have already seen, vowel sounds are voiced. They are produced without any obstruction in the oral cavity. If the branch to the nasal tract is open, the vowel is said to be *nasalized* (such as the vowels in the French words *bon, sans, faim, etc.*). Vowels can be further divided into so-called pure vowels, which can be produced in isolation with a stationary vocal tract, and *diphthongs,* (such as in the words *say, sow* and *sigh*) where a movement of the articulators (the tongue, lips or jaw) is necessary.

Classically, steady vowels are characterized by the location of the highest point of the tongue, and by whether the lips are considered to be rounded or spread. The location of the highest point of the tongue is described in two dimensions: front versus back and high (or close) versus low (or open). The vowel in *feed* is an example of a high front vowel, and that in *fool* is a high back vowel, while those in *cat* and *cot* are examples of front and back open vowels respectively. Front vowels with lip rounding occur in French (*e.g.* in *lune, jeu* and *heure*) and in German (*e.g. in kühl* and *schön*); but they are absent in many languages, including English, Italian and Spanish. Similarly, high back vowels are always accompanied by lip rounding in European languages, though Turkish and several East Asian languages have lip-spread versions.

In contrast to vowels, consonants always involve a narrowing in the oral tract. At one extreme, the narrowing may result in total obstruction. Sounds involving such total obstruction come under the general heading of stops, though the term encompasses two distinctly different sets of sounds. If the nasal branch is open, voiced excitation produces nasal consonants such as the final sounds in *sim, sin* and *sing*. If the nasal branch is closed, no air can flow from the lungs. Pressure builds up, and when the oral closure is released, the resulting turbulent airflow produces a *plosive* consonant. Examples of voiceless plosives occur at the beginnings of the words *pin, tin* and *kin*, while examples of corresponding voiced stops occur in *bun, done* and *gun*. When voiceless plosives are followed by a vowel or other voiced sound, voicing must begin at some instant after the release of the closure. In most dialects of English, the turbulence caused by the narrow opening just after release of the closure is followed by a period of about 100ms of airflow through the larynx with light turbulence. This is known as *aspiration*, and resembles the initial sounds in words like *hop, hip* and *hat*. On the other hand, in most dialects of French vocal cord activity in voiceless stops begins at an instant close to the release, without an intervening period of aspiration. In voiced plosives, vocal cord activity can begin at the instant of release or during the closure as pressure in the oral cavity is built up. Onset of voicing in voiced plosives again tends to occur earlier in French than in English. Although there is no airflow through the lips, some low frequency sound escapes through the walls of the vocal tract when there is voicing during closure.

As we have seen, airflow through a constriction causes turbulence. When this process is steady, the resulting sound is known as a *fricative*, either voiceless (as in the initial sounds of *fat, sip* and *thick*) or voiced (as in the corresponding sounds in *vat, zip* and *the*). The noise-like component of voiced fricatives is generally much weaker than that in their voiceless homologues. Indeed, the whole sound is less intense, and this intensity difference forms one of the cues used in discriminating between voiced and voiceless fricatives.

When the vocal tract is narrowed but not enough to cause turbulence a class of consonant sounds such as the initial sounds in *way, ray* and *lay* is produced. They are lumped together under the general heading of *sonorants*.

This survey of speech sounds is incomplete even for English, but it covers the main categories. The next phonetic unit up from them is the *syllable*

A syllable usually contains one steady vowel or diphthong, though English permits some weak syllables, such as the second syllables in *little* and *station*, that may be pronounced with a so-called vocalic consonant rather than a vowel. The center of a syllable generally corresponds to a peak in acoustic energy, and syllables can usually be located in a speech waveform in this way.

The vowel in a syllable may be preceded and followed by consonants. English, and indeed Indo-European languages in general, especially Germanic and Slav languages, are characterized by the occurrence of strings of consonants (the very word *strings* illustrates the point). Many other languages avoid sequences of two or more consonants. Syllables with no consonants after the vowel are called *open* syllables, and many languages, including Japanese, Chinese and Italian, either permit only open syllables, or at least strongly favor them.

The sequences of consonants surrounding a vowel in a syllable are not arbitrary. Voicing, for example, can be turned on and off only once within a syllable. While some consonant sequences would be physically impossible to pronounce, others are ruled out by the conventions of the language. Thus in syllable-initial position English can have an "s" sound before several other consonants (*e.g.* in *sleep, stick, snow* and *smear*) but not the "sh" sound occurring in *sheep*. By contrast, German permits these same consonants to be preceded by "sh" but not by "s." Some Yiddish (a dialect of German) loan words in English (*e.g. shlepp, schmaltz, shnozzle*) follow the German convention.

Syllables within polysyllabic words may differ in the degree of vocal effort with which they are produced (it is greater, for example, in the first syllable of *after* and in the second syllable of *before*). Such *stress* differences manifest themselves as differences in the loudness, duration and pitch of the syllable, and sometimes in the quality of the vowel, which tends to be more central in position in unstressed syllables. The location of stressed syllables is regular in some languages, always occurring, for instance, on the first syllable in Czech and usually on the penultimate syllable in Italian. The rules for stress location in English are complex, while syllables in French have little or no stress differences between them.

We can now go on to look briefly at the acoustics of speech production.

Whether the excitation in a speech sound is voiced or voiceless, the acoustic signal generated by the excitation is modified by the resonant structure of the vocal tract, which behaves as an acoustic tube along which planar propagation of sound waves occurs. Differences in the cross-sectional area along the length of the tube cause reflections, and it is these reflections that give rise to the resonances or *formants*. The resonant structure therefore depends on the position that the tongue, lips and jaw are in.

The generation of the excitation and its spectral modification by the vocal tract turn out to be largely independent of each other. To a good approximation, they can therefore be considered as a source isolated from, and leading into, a linear filter.

48

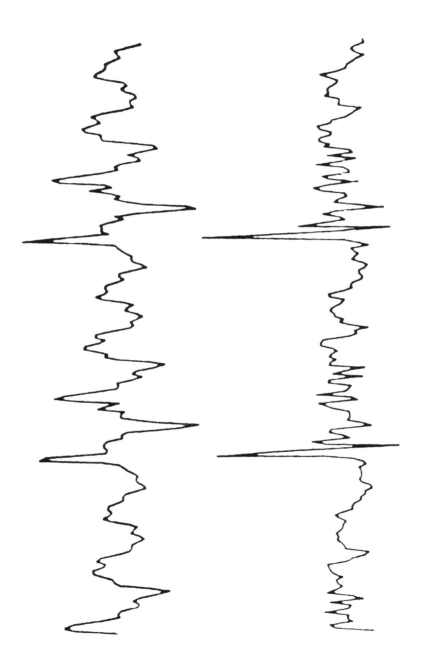

Figure 1. Upper trace: a 20ms portion of the time-differenced waveform of a neutral vowel produced by a male speaker. Lower trace: the same waveform with the effect of the vocal tract removed.

The upper trace of Figure 1 shows a 20ms stretch of the waveform of a non-nasalized vowel (strictly, it is the *time-differenced* waveform: differentiation provides a 6db per octave lift, which serves to flatten the long-term spectrum for voiced speech). Notice that the waveform consists of a pattern that repeats itself at regular intervals. The repetition rate is the rate at which the vocal cords come together — the *fundamental frequency* of this speech sound — while the repeating pattern itself is the response of the vocal tract to this periodic excitation.

The lower trace in Figure 1 shows the excitation with the effect of the vocal tract removed. The impulse-like excitation occurs each time the vocal cords come together and close off the airflow from the lungs. In the particularly simple vowel shown here (the central, "neutral" vowel occurring in a word such as the standard British English pronunciation of *bird*) the impulse travels from the larynx to the lips, where part of it is radiated into the open air beyond and part is reflected back towards the larynx with its polarity reversed. At the larynx the signal is reflected again, this time without polarity reversal, and it continues to bounce between larynx and lips steadily losing energy by absorption in the walls of the vocal tract, by absorption below the vocal cords, and by radiation to the outside world, until the next excitation impulse comes along. The pattern of an impulse emerging with alternating polarity can be seen in the upper trace of Figure 1. The impulse gets rounder as time progresses because high frequency components are lost faster than low frequency components.

Figure 2 shows the power spectrum of a section of speech waveform like the one in Figure 1. The regularly spaced spikes occur at each integer multiple of the fundamental frequency of the excitation, and are *harmonics* of the fundamental. The intensity of the harmonics is determined by the product of two factors. The first is the spectrum resulting from the details of the airflow through the larynx from one closure of the vocal cords to the next; and the second is the spectrum corresponding to the impulse response of the vocal tract.

Let us look at the laryngial component of the spectrum first. This component is generally smooth, and above a few hundred Hertz it declines at about 12 dB per octave. To some extent, however, this decline is counterbalanced by a 6 dB per octave *rise* due to the effect of radiation from the mouth, giving a net decline of the excitation spectrum of voiced speech of around 6 dB per octave. An impulse has a flat power spectrum, so in order to make the excitation signal impulse-like its spectrum must be made roughly flat. This is why Figures 1 and 2 used differentiated speech. The impulse response of the vocal tract then appears directly in the upper trace of Figure 1.

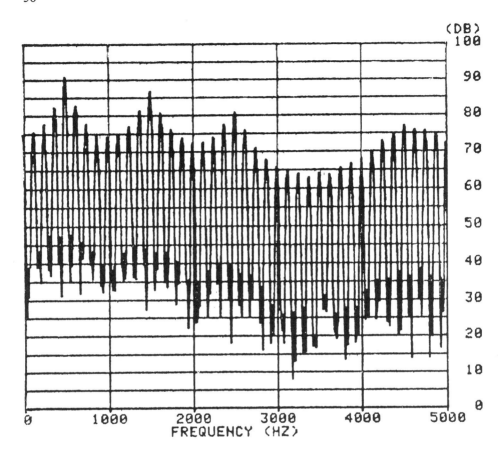

Figure 2. The power spectrum of a time-differenced neutral vowel.

The exact shape of the voiced excitation spectrum varies from individual to individual and changes with the intensity of the speech and the mood of the speaker. A louder or tenser voice tends to be associated with the vocal cords being open for a shorter proportion of the excitation cycle and with their closing more rapidly. This results in the excitation spectrum declining more slowly. Sometimes, the vocal cords open and close periodically but closure is incomplete, causing turbulent flow through the remaining small gap and a *breathy* voice quality. This occurs for most English speakers in the "h" sound of words like *behind* and *mayhem,* though some speakers have this kind of excitation for all their voiced speech. Other speakers, particularly at the ends of sentences have a *creaky* voice quality, in which the excitation period can become irregular, often alternating between long and short periods

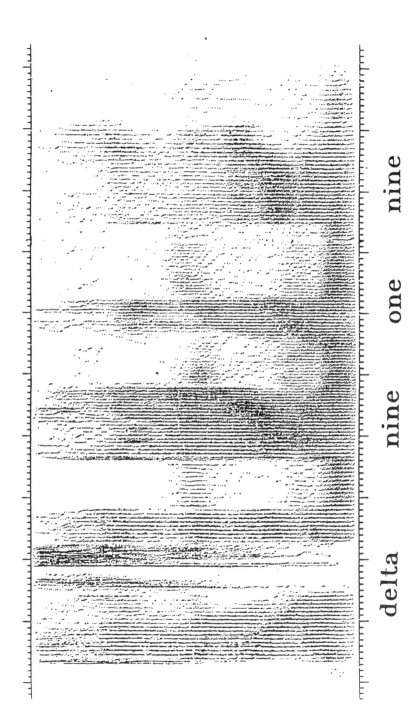

Figure 3. Spectrogram of the word sequence "delta nine one nine" produced by a male speaker.

and often showing significant excitation on the opening of the vocal cords as well as on their closure [1]. In English, and indeed in most languages, however, such variations in the excitation signal are not used to carry information about the explicit content of a spoken utterance. Incidentally, since women have "higher pitched" voices than men, they are often assumed to have more intense high frequency components in their voices. If anything, the reverse is true: women generally have higher fundamental frequencies, so the harmonics are further apart, but the excitation spectrum tends to fall off more rapidly in women's voices than in men's.

The intensities of the harmonics in Figure 2 show a series of smooth peaks. This structure is due to the impulse response of the vocal tract, and the peaks correspond to formants. Formants are numbered in order of increasing frequency. In the particularly simple sound illustrated in Figure 1 the vocal tract resembles a tube of uniform cross-sectional area from the vocal cords to the lips. For a typical male vocal tract, this gives rise to a first formant at about 500 Hz and to subsequent formants spaced 1 kHz apart at 1.5 kHz, 2.5 kHz, *etc*. Since the vocal tract of a woman is typically 10 to 15% shorter, the corresponding formant frequencies are raised by this amount.

Figure 3 shows a *spectrogram* of the sequence of words "delta nine one nine." The horizontal axis corresponds to time and the vertical axis to frequency from 0 to 5 kHz. Regions of high energy appear dark. Since the analysis used here has a lower frequency resolution than that in Figure 2, harmonics of the fundamental are not resolved. The vertical striations correspond to the excitations caused by the vocal cords, while the broad horizontal or sloping bars are formants.

Figure 4 shows the second formant being excited as the airflow through the vocal cords is stopped. The loss in energy after excitation is roughly exponential, though the rate of energy loss is greater when the vocal cords are open than when they are closed, since energy is absorbed into the trachea and lungs during the open phase. The increased damping during this phase also causes a slight decrease in the frequencies of the formants. As we have seen, high frequency energy is lost faster than low frequency energy. Consequently, the higher formants have larger bandwidths than the lower ones.

The excitation in voiceless sounds resulting from turbulence in the vocal tract resembles white noise. As with voiced sounds, however, radiation effects from the lips tend to reduce the intensity of the low frequency components, and in voiceless sounds energy in the first few hundred Hertz is consequently weak. In voiceless sounds, the first formant is not normally excited. Formant structure is much less marked or even — particularly for "f" sounds — non-existent.

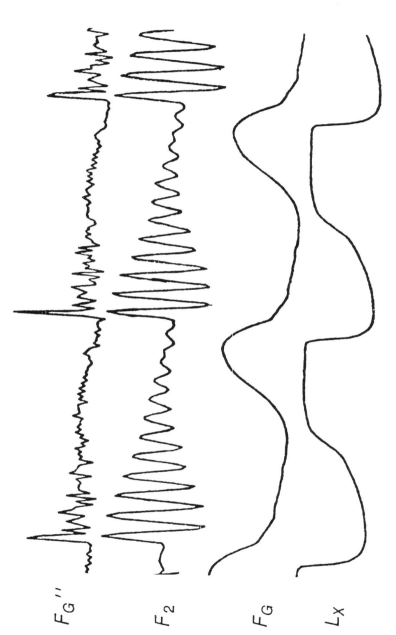

Figure 4. 20 ms of voiced speech from a male speaker. The bottom trace, labeled L_X, is a measure of the electrical impedance across the larynx and correlates strongly with the area of contact of the vocal cords, with increasing contact being in the downward direction on the trace. The trace above it, labeled F_G, shows the airflow through the vocal cords. The next trace up, labeled F_2, shows the waveform corresponding to the second formant only. Finally, the top trace, labeled F_G'', shows the impulse-like waveform produced by differentiating F_G twice. This corrresponds to the lower trace in Figure 1.

The description of voiced speech in terms of an impulse response and the frequency of the impulses has several advantages. The impulse response varies as the positions of the tongue, jaw and lips are changed, while the fundamental frequency depends on the muscles that control the tension in the vocal cords and on the air pressure behind the vocal cords. For the most part, changes in the settings of the larynx and vocal tract occur slowly relative to the perceptually important frequencies in the speech waveform, which are determined by the time between successive reflections of sound waves in the vocal tract. Thus, while we need to sample the speech waveform at least eight thousand times a second to obtain a reasonable digital representation, a description in terms of fundamental frequency and a few parameters describing the impulse response typically needs to be updated as little as a hundred or even fifty times a second, and even then the changes between updates tend to be small.

A second major advantage of an impulse-response/fundamental-frequency description is that the two factors perform separate linguistic functions. In most western languages the identity of a word does not depend on the fundamental frequency pattern with which it is uttered. In some other languages, such as Chinese, the identity of a word may depend on the fundamental frequency pattern, but even then an analysis strategy must still separate the two factors: the fundamental frequency pattern and the configuration of the articulators in the vocal tract remain substantially independent attributes of the word.

For non-nasalized vowels and some non-nasal consonants the impulse response of the vocal tract is quite accurately modeled by a set of resonances in series; that is, the vocal tract can be regarded as an all-pole filter, and its effect can be completely specified by the frequencies and bandwidths of the poles, corresponding to formants. For such sounds, a technique known as *linear predictive coding* (LPC) can in principle be used to determine from the waveform the frequencies and bandwidths of the resonances (see the book by Markel and Gray [2]).

In other sounds, notably in nasal consonants and nasalized vowels, the all-pole model of the vocal tract is not valid. Resonances are configured in parallel as well as in series, and consequently zeroes, corresponding to antiresonances, appear in addition to poles in the transfer function of the vocal tract filter.

THE PERCEPTION OF SPEECH AND OTHER SOUNDS

Human hearing is similar to that of closely related animals, who, of course, do not use speech. Therefore, it does not seem to have adapted to the properties of the speech signal. Rather, speech must have evolved to suit the properties of our sense of hearing. In artificially generating speech or in trying to transmit it efficiently, there is clearly no point in striving to reproduce features that are inaudible. Equally, in speech recognition it would be misguided to depend on features that are inaudible to humans, since a speaker is unlikely to control features that he or she cannot hear, unless they are locked to other, audible, features, in which case they carry no additional information. It is therefore important to take account of our, admittedly limited, knowledge of human hearing.

Our impression of the loudness of a sound corresponds better to the log of the acoustic power rather than its linear value. Thus, successively doubling the power in a sound gives an impression of roughly equal steps in loudness. The loudness of a sound is therefore normally expressed on the logarithmic *decibel* (dB) scale, where a factor of ten increase in power corresponds to 10 dB. (The cube root of the power is now generally considered to match perceptual loudness even more closely, but the use of the log scale has been retained.)

The amplitude sensitivity of our hearing peaks in the 1 to 2 kHz range. It falls off markedly somewhere below 100 Hz and, depending on our age, somewhere above 5 to 10 kHz.

The frequency sensitivity of the ear can be measured in various ways — by having listeners determine subjectively equal frequency intervals at different locations in the spectrum; by testing their ability to detect small changes in frequency; by measuring the frequency range over which spectral components interact; or even by direct physiological measurements on the inner ear. All these methods lead to strikingly similar perceptual frequency scales, with sensitivity being roughly constant over the first few hundred Hertz and then decreasing with increasing frequency. The perceptual frequency scale is often approximated by a scale, the *technical mel scale*, that is linear to 1 kHz and logarithmic from then on.

Just as one might expect from signal processing considerations, the degradation in frequency resolution at higher frequencies is associated with an improvement in temporal resolution. This trade-off is well matched to the acoustic properties of speech. As we saw in Section 2, the higher formants have large bandwidths and do not therefore require high frequency resolution. In voiceless sounds, energy tends to be concentrated at high frequencies. Spectral fine structure is absent, but such sounds, particularly voiceless plosives, contain events that are sharply defined in time. Voiced sounds therefore require good frequency resolution at low frequencies (below 2 kHz) and voiceless sounds require good temporal resolution above 2 kHz.

Unless two frequency components are within a certain critical distance of each other on a perceptual frequency scale, their phase relationship has no perceptual effect. Consequently, a sound can be substantially characterized by its power spectrum, ignoring its phase spectrum.

Strong frequency components can suppress the ear's response to weaker components. In *temporal masking*, the strong component masks a weaker component at the same or a nearby frequency. The stronger component can occur just before or just after the weaker component, though the effect operates over much greater temporal separations in the former case — so-called *forward masking* — than in the latter. In *simultaneous masking* or *frequency masking* a strong component masks the presence of a weaker component presented at the same time at a different frequency. The effect decreases as the frequency separation between the components increases, but the decrease is slower when the weaker component lies above rather than below the stronger component. Frequency masking therefore operates primarily upwards in frequency.

The use of our two ears allows us to deduce the direction of a sound source in the horizontal plane, since there will generally be a difference between the time of arrival of an acoustic event at the ears that depends on the direction from which it is coming. In addition, the shape of the external ear appears to have a direction-dependent filtering effect on sounds that allows some directional sensitivity in the vertical plane and discrimination between sound sources behind and in front of the listener. These capacities certainly contribute to our ability to follow a particular conversation in a crowded room, though this ability also seems to exploit a more sophisticated mechanism that allows us to track a particular voice.

So far, we have been considering the perception of sounds in general. Let us now turn to consider speech sounds in particular.

Klatt [3] showed that listeners use different criteria when judging the *phonetic* similarity of two speech sounds from those they use when simply judging the acoustic similarity of two sounds. For example, changes in the spectral balance of the signal such as are caused by manipulating the tone controls on a stereo have little effect on phonetic judgments. This makes us immune to the spectral tilt effects of the telephone, of room acoustics and of shouting.

Phonetic judgments in voiced speech turn out to depend strongly on the frequencies of the first three formants, though not on their bandwidths, nor on the details of higher formants. This sensitivity to the lower frequency and most intense peaks in the spectrum can perhaps at least partly be explained by simultaneous masking, which would tend to mask the weaker higher formants and spectral details in the regions between the lower formants.

At this point it might be interesting to look at the extent to which two analysis techniques, LPC and mel-scale filter banks, that have both been widely used in speech recognition and in speech transmission incorporate the perceptual properties discussed so far. Both represent the short term power spectrum on a log scale, ignoring the phase spectrum. If LPC is viewed as a technique for matching the power spectrum, it has the interesting property of not making a least-squares fit to the whole spectrum as one might expect but rather of concentrating on fitting the strong parts — *i.e.* the formant peaks — well. On the other hand, conventional LPC is unable to reflect directly the non-uniform frequency resolution of the ear. A filter bank can simply reflect perceptual frequency resolution in the width and spacing of its channels.

When the vocal tract is reopened during a plosive sound the formant frequencies pass through rapid transitions as the articulators involved in the closure move apart. Our hearing system is particularly sensitive to these formant transitions, and they constitute strong cues to the identities of plosives.

By manipulating such transitions in synthetically generated speech stimuli, the boundaries between speech sounds — between "b" and "d" sounds, for example — have been probed. It turns out that consonant sounds are perceived categorically [4]. That is, sounds are not perceived as partly "b-like" and partly "d-like;" rather, they are perceived as fully either one or the other. Any such effect in the perception of vowels is much less marked.

We saw in the previous section that a production-oriented approach can lead to an efficient description of the speech signal because the articulators involved in speech production move slowly relative to the time between successive reflections of sound waves in the vocal tract. If the motion of the articulators could be derived directly from the speech waveform, it might provide a particularly good representation for the perception of speech. The Motor Theory of Speech Perception [5] holds that this is exactly what human listeners do. Although the more extreme expressions of this view are probably less popular now than they once were, it must surely have some validity. In fluent speech the articulators rarely reach the extreme positions occurring in speech sounds spoken in isolation; rather they take short-cuts between the positions needed for neighboring sounds, and the degree of the short-cuts depends on the carefulness and rate of the speech. It is hard to imagine how a speech perception mechanism could handle the acoustic variations caused by this behavior without resorting to a model of speech production.

Automatic speech recognition, in particular, would undoubtedly be helped enormously by a thorough understanding of how humans routinely accomplish the task. Sadly, though, we are still far from such an understanding. Some of the points discussed in the next section may make the magnitude of the problem a little clearer.

SPEECH AS A COMMUNICATIONS SIGNAL

Speech is a signal with an intended message. In this respect it differs from, say, EEG signals or from a signal transmitted from a satellite representing an image of a portion of the earth. Such an image has the obvious difference that it is two-dimensional while the speech signal is effectively one-dimensional. The more important difference, though, is that the satellite image is not a communication: it contains information but it does not contain a message. The very same image might be used to study the vegetation of an area or to try to spot missile silos, but presumably the image processing techniques appropriate for the one task would be quite different from those appropriate for the other. Thus, image processing tends to be a loose collection of techniques with diverse goals.

Apart from certain applications such as speaker recognition, speech processing is concerned with the intended message: with transmitting it, recognizing it, or generating it. It is, therefore, a narrower, more focussed activity than image processing.

The discussion that follows excludes certain kinds of social communication such as "Hello, how are you?," where the speaker is not so much enquiring into the state of health of the listener as making a semi-voluntary announcement of his or her feelings and relationship to the listener. This use of speech is similar to the way in which a dog might bark a greeting at its owner or a threat at an intruder. It is not what makes human speech special, and it is not of primary interest in communicating with machines or, presumably, in military communications.

Speech communication can be usefully compared with man-made artificial communications signals, such as H.F. teleprinter transmissions or telephone dialing signals. In such signals, there is quite clearly a message, and the message is laid out sequentially in time or space just like speech. The similarities to speech are obvious; the differences much less so, but they are nonetheless large and worth studying.

The artificial signals in our examples are composed of a sequence of units, the units being selected from a definite, known set that we could call an *alphabet*. The units in a message are generally well separated from each other, and they do not interact (Figure 5). The decoding device usually has available to it in some form an *ideal*, undistorted representation of the alphabet, and decoding consists mainly of trying to identify the received units one by one using its built-in knowledge of the ideal forms.

What is the equivalent of these units for the speech signal? There seems to be there is no single exact equivalent. Perhaps the closest candidate is the word, but words differ in several major respects from our artificial units.

First of all — notwithstanding our prejudices from the written form of language — spoken words do not in general have gaps between them (see Figure 3, where the only gap occurs before the "t" in "delta"). Indeed, there are no consistent acoustic cues of any kind to word boundaries. What is more, not only are words not well separated from each other, they often interact at their boundaries. For instance, "bread board" is often pronounced in fluent English in a way that we might write as "breab board," and "this shop" as "thish shop." Indeed, in fluent speech, short, low-content words such as *the, of* and *a* are so strongly influenced by their context that they are often unrecognizable when excised from it.

Next, we know of no ideal reference forms of words: any normally pronounced version of a word is as good as any other, and no two productions will ever be exactly the same. In particular, words differ in their *prosodic* features (intonation, timing and loudness) depending on their function in a sentence. Even in such a prosaic utterance as a list of digits, the final digit differs markedly from the others, being typically 60% longer and having a falling intonation (see Figure 3). When people try to generate synthetic sentences by recording words in isolation and playing them back unmodified in a sequence, the result is disastrous — each word is perfectly clear, but it is almost impossible to grasp the meaning of the sentence.

When words were suggested above as the best equivalent of artificial communication units, some readers may have been surprised that *phonemes* were not proposed. Such surprise would be understandable considering the number of popular articles on speech technology that talk about speech being made up of phonemes as though it were like laying out bricks in a line — just like the symbols in teleprinter transmissions. Proponents of phonemes might also point out that the phoneme inventory (just over forty in English) is much more manageable — more alphabet sized — than the enormous inventory of words in a language. Some people might also be influenced by the way words are printed as a string of discrete context-independent letters. Despite all this, phonemes bear little resemblance to teleprinter symbols. If we must have a writing analogue for phoneme sequences, quite a good one is provided by hastily scribbled handwriting, in which individual letters are hard to isolate and depend for their form on the other letters around them.

A phoneme is defined as *the smallest unit of speech within a word that when changed results in a change in the meaning of the word.* Thus, the English word *tap* differs from the English word *cap* in the position of the tongue at the start of the two words. In *tap* the point of contact between the tongue and the roof of the mouth is just behind the upper teeth, while in *cap* it is towards the back of the mouth. We can conclude that *cap* and *tap* must start with a different phoneme. We could have started with the tongue making contact in other places: it could have been directly behind the upper teeth like the "t" sound in *eighth*, or the tip of the tongue could have been curled back slightly like the "t" in *tree*. If we used either of these "t" sounds in our word *tap* we would *not* get a new word, we would simply have *tap* with a slightly non-standard pronunciation — we might not even notice that the word sounded odd if it occurred in fluent speech. Yet those same "t" sounds represent different phonemes for some other languages. For speakers of such languages (which include several major languages spoken in India) the "t" variants presumably sound quite distinct. In the same way, the English "l" and "r" sounds in words like *lap* and *rap*, which sound quite different to English speakers, do not correspond to different phonemes in Japanese, so Japanese speakers have difficulty in making the distinction.

Thus, cues that provide phonemic distinctions in a language are much more noticeable than those that do not. English is often considered not to have nasalized vowels, but in fact they are as common in English as in French; it is simply that nasalization is not phonemic in English: that is, nasalization cannot change the meaning of a word, and its presence is optional. While the vowel in French *canne* is never nasalized, that in English *can* almost always is, though we probably would not notice if it were not. Failing to nasalize a French nasal vowel is very noticeable: it produces either nonsense or a different word; for instance, *baton* (stick) would turn into *bateau* (boat) in standard French if the nasalization were removed.

Phonemes, then, are not "speech sounds" in some absolute sense, they are a property of the way a language gets coded in sound, and their phonetic realization is frequently context dependent. Something interesting is happening in standard French right now: the vowel sounds in the words *jet* (jet) and *gel* (frost) used to be different phonemes, that is to say, there existed pairs of words such as *pré* (meadow) and *près* (near) that differed just by the fact that the first had the *jet* vowel in it and the second the *gel* vowel. French speakers are increasingly using a new rule that says that the *jet* vowel can occur only at the end of a word and the *gel* vowel only when followed by a consonant sound. Thus the *pré/près* distinction is lost, and the two vowels have become context-dependent *allophones* of the same phoneme. French has lost a phoneme, but it has *not* lost a speech sound.

So far, we have established that a phoneme does not correspond to a single speech sound, but perhaps we could say that it corresponds to a set of sounds. If by "sounds" we mean something we can hear and identify in isolation, the answer has to be no, or at least not always. The English word *dell* (or the first syllable of *delta* in Figure 3) is made up of three phonemes /d/ and /e/ and /l/ (phonemes are conventionally written between oblique lines); but as Figure 3 shows the syllable consists of a continuous acoustic sequence, and there is no way of pronouncing the /d/ without also pronouncing the vowel after it. If we take a recording of *dell* and listen to what happens as we successively chop off more and more of the end, we never get to hear a /d/ in isolation: when we have shortened it enough that we no longer hear the vowel, we no longer hear anything that we perceive as speech.

Vowels, of course, can be produced and perceived in isolation. But in the *dell* example just described, when the word has been shortened to the point where the "l" sound is no longer heard, the vowel is not perceived as the "e" in *dell* but as the reduced (*schwa*) sound heard in an unstressed *the*.

The picture of what a phoneme might be in acoustic terms gets even fuzzier when we start to ask about the acoustic features a listener might use to decide what phoneme sequence he or she is hearing. By using a speech synthesizer, researchers have been able to vary the properties of speechlike sounds and so investigate the phonetic cues that listeners use. It turns out that they often do not depend on a single cue but rather weigh the evidence from several independent features. Some results have been particularly surprising. For example, the words *ones* and *once* are normally felt to differ just in their last phoneme, *ones* ending in the voiced phoneme /z/ and *once* in the corresponding voiceless phoneme /s/; but it is possible to change a listener's judgment of which word is being presented merely by altering the length of the /n/ sound (a longer /n/ causing *ones* to be heard). Indeed, this is probably the most important phonetic cue in discriminating between these words in natural speech. Here we have an example, then, where the major distinguishing mark of a phoneme is not only not what we would expect it to be, it is not even *where* we would expect to find it.

Moreover, cues to phoneme identity are not even entirely confined to the auditory channel: in appropriate circumstances visual cues can be integrated into speech perception. The point has been convincingly demonstrated [6] by synchronizing a recording of a plosive-vowel sequence — e.g. "ba" — with a video recording of a person producing a different stop consonant followed by the same vowel — e.g. "ga." The perception of the sound is strongly modified by the conflicting visual cues — in the ba/ga example what is perceived is "da." The effect has perhaps to be seen to be fully believed: when I saw a demonstration I "heard" a perfectly natural "da" as long as I watched the screen; as soon as I looked away it reverted to "ba."

Speech, then, clearly cannot be considered as a simple sequence of speech sounds, nor even as a sequence of discrete words. At the acoustic level, there are no known discrete units whose ideal forms can be defined. This is a further reason why, in contrast to artificial signals, it is useful to study the production of the speech signal in order to describe its acoustic properties.

The fact remains, however, that we do have a strong internal impression of speech as being made up of neat sequences of words and words as being made up of neat sequences of discrete, context-independent speech sounds.

Visual perception may provide a clue to what is going on in speech perception. Information on a scene reaches us as a two-dimensional pattern of light on our retinas, yet we perceive a world of three-dimensional objects. This process is not strictly dependent on stereo imaging or on the lens adjustment needed for focusing, because we have no difficulty in interpreting scenes on a television or cinema screen, where such information is absent. Our visual perception is not a passive reception of a pattern of light but rather an *active reconstruction* of a scene based on the visual evidence and on our knowledge of the world. People who have been blind from birth and who gain visual function as adults are said to have great difficulty in learning to see; even though the information transmitted by their optic nerves may be the same as that of other sighted people, they have simply not learned to interpret that information. In normal individuals this interpretation is unconscious and cannot be turned off. When we look at a drawing or a painting of a scene we automatically interpret it in three dimensions. If the picture contains paradoxes that prevent a consistent three-dimensional interpretation, such as, for example, in many of the works of the artist M.C. Escher, we cannot choose to avoid the paradox by perceiving the picture as a meaningless pattern of light and dark on a flat piece of paper. Instead, we are compelled to go on trying to "make sense" of it as a three-dimensional scene.

Just as our visual system does not function like a camera passively recording incident light, so our perception of speech cannot be likened to the action of a microphone passively transcribing acoustic signals. Rather, we actively reconstruct the message from the various phonetic and prosodic cues in the signal together with our knowledge of the vocabulary and syntax of the language, of what would be meaningful and germane to the situation, and of the known habits of the speaker.

This reconstruction process is so effective and automatic that we are not normally aware that it is going on. On the telephone, for example, we rarely notice that the final "s," "th" and "f" sounds of *lass, lath* and *laugh* are virtually indistinguishable; it is only when we have to note down an unfamiliar name that we become aware of just how much acoustic information is missing.

Even when the acoustic signal is undegraded, our perception of speech sounds can be switched by information from other parts of the sentence. Thus, when we are primed with *the dogs chased the* **cats,**

we tend to hear the completion of the sentence as

and the **cats** *shinned up the tree;*

whereas if we are primed with

the dog chased the **cat,**

we tend to hear

and the **cat** *shinned up the tree,*

even though the second half of the sentence would be pronounced identically in the two cases.

It is the reconstruction of a spoken message that gives us such a firm impression that the speech signal consists of a neat sequence of phonemes: it may indeed be possible to describe speech in this way, but only at a certain stage of processing in our brains, not at the level of the acoustic signal.

The information used in reconstructing a spoken message can be drawn from many different levels and uses both acoustic information and the listener's knowledge. We have already noted the existence of phonetic cues, which indicate word structure, and prosodic cues, which generally indicate sentence structure and point to the location of significant information in the sentence. In addition to the rules that govern the order in which words can be uttered in the syntax of a language, there are also agreement rules, such as those between a verb and its subject and between an adjective and noun it qualifies. Though limited in English, such rules can provide much disambiguating information in other languages. Gender distinctions, for example, can operate like a check bit in a coding scheme: the French words *boisson* (drink) and *poisson* (fish) are acoustically similar, but since they differ in gender, the phrases *une boisson délicieuse* (a delicious drink) and *un poisson délicieux* (a delicious fish) are much more distinguishable. A listener might also apply expectations that a sentence should be meaningful and germane to the situation. Finally, the work with synchronized video recordings demonstrates that in some circumstances optical information is used in reconstructing the speech message.

The amount of external cues needed for effective reconstruction depends on the predictability of the message: a mere grunt might be perceived as *Merry Christmas* on December 25'th; but if, for some reason, one wanted to greet someone in that way in mid-summer, the words would have to be very clearly articulated.

The list of different sources of information that can be used in decoding a spoken message points up another way in which speech differs from the teleprinter transmission, namely, the fact that speech has to be regarded as a *multilevel* sequence. Thus, words can be thought of as phoneme sequences, while they themselves form part of word sequences making up phrases, which in turn make up sentences. Evidence needed to understand speech is present at every level, and the evidence at all levels probably has to be considered simultaneously if the message is to be understood. It is true that we could find much the same set of levels in a teleprinter transmission of meaningful text, but the levels are not so intimately mixed: in order to decode the individual teleprinter symbols we do not even need to know what language the text is written in.

Speech is often said to be a redundant signal. It is argued that the same utterance can be understood either when it is low-pass filtered at 1 kHz or when it is high-pass filtered at 1 kHz, so the information below 1 kHz must be duplicating the information above that frequency. This reasoning is faulty. The amount of information one needs in a speech signal depends on how skilled one is at reconstructing the message: much more acoustic information is needed when the topic of the message is obscure or when a language is being used that is not the native language of the listener, even though all the words and constructions may be familiar. Native speakers presumably have better information on the relative probabilities of words and constructions. What may be more important, they also know which constructions are not allowed in the language, while non-native speakers cannot distinguish between impossible constructions and constructions that are unfamiliar to them but that are nevertheless possible. As a result, the non-native speaker may waste valuable processing effort on the pursuit of hypotheses that a native speaker would not even consider, just as chess masters are said not to see bad moves.

There are tradeoffs, then, between the information available in the listener's brain and the information needed in the signal itself. Speakers apparently take this tradeoff into account and adjust the amount of acoustic information they provide for each word in their speech in the light of their subconscious estimates of the predictability of these words [7,8]. For instance, working with phrases in Swedish, Hunnicutt found that the word corresponding to "the letters" excised from the spoken phrase "During the morning the postman quickly delivered *the letters* which had collected during the weekend," where its occurrence is predictable from the context is less easily recognized when presented in isolation than when the word is excised from the following context in which it is less predictable: "Curiously the man examined *the letters* which he had found."

Speakers, then, do not emit speech messages to be picked up by anyone who cares to listen, they *talk to someone*. Although we as yet know too little about speech to be sure about this, it seems likely that a speaker puts just enough cues into the speech to allow the listener (or imagined listener in the case of, say, a radio broadcast) to be able to comfortably reconstruct the message from the evidence available. Thus, when we are saying something that is difficult to follow, or when we are speaking to someone we believe to be foreign, deaf or senile, we supply more phonetic information than we would in a relaxed conversation with a friend. Elision of phonetic information, such as when we say *fish 'n chips,* is often ascribed to laziness, but it can be seen to be a rational strategy for the economical use of a communications link: it would be lazy only if the person at the other end of the link were obliged to make an unreasonable effort to reconstruct the message. Depending on the circumstances, overarticulation can be just as inappropriate as underarticulation: it can sound stilted, irritating, even insulting when the listener feels it to be unnecessary.

To summarize this section: the speech signal is different in nature both from messageless signals such as satellite images and from machine-generated message-bearing signals like the teleprinter transmission. It is a signal from which a message may be reconstructed using information drawn from many sources, both information at various levels in the signal itself and information stored in the mind of the listener. The amount of information that the speaker puts into the signal depends on the difficulty that he imagines the listener will have in reconstructing the message from it.

SPEECH AND WRITING

As a species, we developed speech long before we developed writing. As individuals, we learn to speak before we learn to write, and speech remains for most of us our primary means of communication with each other.

It may seem surprising, then, that when we think about verbal communication our image is drawn almost entirely from *written* communication. But text is literally easier to visualize than speech. Unlike ephemeral speech, text stays on the page to be examined. At school, our assignments and examinations are overwhelmingly in written form. We become conscious of the rules governing written language and skilled in applying them. We come to regard everyday spoken communication — if we think about it at all — as an inferior version of the written language, a version lacking in elegance and littered with errors. At a conscious level, at least, we tend to ignore those features of spoken language, such as prosody, that are not represented in the written form.

We saw in the previous section that printed text with its discrete, context-independent letters and words can incite a false impression of what speech is like. Printed text probably reflects an internal representation of speech after much sophisticated processing has been applied to it. But the cultural importance of printed text has meant that the properties of text have been projected back onto speech, reinforcing the belief that our internal impression of the speech signal corresponds to an external reality.

A similar phenomenon occurs when we think about the style of language appropriate for speech. Yet even though the formal rules of grammar underlying the two modes of communication are generally thought to be the same, the styles of language appropriate for writing and speaking are different. In terms of these formal rules, spoken language is more errorful, partly because we have much less time to plan and polish our spontaneous speech than we have our writing, though many so-called errors in speech may actually be observances of different rules. Certainly, spontaneous speech with all its apparent ungrammaticality, redundancies, hesitations and incomplete constructions is usually easier to follow than text in written style being read aloud.

Papers delivered at conferences are often all but impossible to follow because the presenter is reading a text written in a style appropriate for a reader but not for a listener. When the presenter departs from his text to comment on a slide or to answer a question he generally becomes much easier to follow, even though his language might appear to be less well formed. Speakers are sometimes tempted to read a prepared text because they believe it allows them to pack more information into a limited time. It does indeed allow them to *transmit* more information, but it does not allow their audience to *receive* more information. The rate at which we transmit information in a well planned talk without a prepared text probably corresponds to the rate at which the audience can absorb it.

Until recently, an example of the inappropriateness of written style in speech could be heard regularly on a U.S. television network when a sponsoring corporation described itself as:

> "... providing high-technology, computer-based systems solutions to the complex problems of business, government and defense."

Admittedly, this is not the snappiest of sentences even as text, but when spoken it is particularly indigestible. Participial phrases like this one are not common in speech. This example contains a large proportion of long, relatively rare, ostensibly high information-content words, while real speech contains more short, common words. Many of these long words act as qualifiers piled up before 'solutions' to an extent that strains our auditory ability to hang in until the noun comes along. When we read the same sentence this problem does not arise.

A similar piling up of subsidiary information that seems to be unacceptable in speech but common in writing occurs when a main clause is preceded by a subordinate qualifying clause starting with *although, while* or *since*. The rarity of such constructions in speech is presumably due to the strain they put on our ability to wait for the subject of the main clause.

The words commonly used to link or separate ideas in spontaneous speech are generally different from those used in text. Words like *moreover, however, nevertheless, thus, therefore, consequently,* and many others are common in text but rare in speech. Simple link words like *and, but* and *so* are more common in speech than in text, and a further set of link words like *well, O.K., right, look, besides* and *anyway,* are common in speech but rare in text.

Partly because text fails to reproduce many of the cues supplied in speech, faithfully transcribed spontaneous spoken dialogue is all but incomprehensible. Stubbs [9] provides a transcription of a conversation between himself and two schoolboys, which — though he assures us it was well ordered and comprehensible to a listener at the time — is almost impossible to follow as a text.

Chapanis [10] has described an experiment that allowed direct comparison between spontaneous speech and written communication. Pairs of subjects who were physically separated from each other were required to carry out a task needing their active cooperation. Performance on the task was compared when various channels of communication were provided. These included a speech link, a teleprinter, a means of passing handwritten notes, and a video link, together with various combinations of these channels. Subjects were able to complete the tasks roughly twice as quickly with any combination that included voice as they could with any combination that did not include voice. Transcriptions of voice communications showed, as one might expect, a high proportion of mal-formed or incomplete sentences, but so did the written communications in those cases where written messages could be passed back and forth without any delay. It seems that failure to observe the formal rules of grammar may be a feature of any spontaneous dialogue rather than specifically of spoken dialogue. Subjects used about five times as many words to solve the same task using a voice link as they did using a text link, but they delivered the spoken words ten times faster than the written ones.

If humans optimize their use of any communications channel, then these results, and the differences in spoken and written style noted earlier, are consistent with the idea that writing or reading a single word is relatively more expensive in time or effort compared with speaking or hearing a word, but recognizing spoken words is a more uncertain process than recognizing written words. In particular, the occurrence of unusual or unlikely words or of complex sentence structures poses more of a problem for a listener than for a reader. The additional uncertainty in decoding speech has thus to be countered by using more words and by using simpler words and simpler constructions; but this need to use more words is offset by the greater speed with which spoken words can be delivered.

Printed text is either legible or illegible. On the other hand, when we first hear a speaker we have to adapt to the peculiarities of the voice. If the speech is unusual, if the speaker has a foreign accent, for example, or if the acoustic signal is degraded, as it is on the telephone, there is a noticeable period over which we need to be gathering information on the characteristics of the speech we are hearing. Consequently, the most effective speech to use at this point is the most predictable speech, since anything that is not predictable will not be understood and will be less useful in supplying the information that the listener needs. The influence of text, however, may lead us to ignore the need for this adaptation period. For example, the *Office de la Langue Française* of the Government of Quebec produces a booklet [11] giving advice on the use of the telephone in French. They recommend that private individuals should answer the phone by saying just *Allô!*, rather than *Allô! j'écoute*. In the case of the formula to be used by receptionists or telephonists in answering the phone on behalf of an organization, they recommend giving just the name of the organization. In both cases, they assert that adding *bonjour*

"is not only useless but also incorrect." [translation].

This advice reflects influence from written language, where good style requires the number of words used to be minimized, as opposed to spoken language, where additional words cost little in time or effort and where predictable words like *bonjour* or *j'écoute* can provide useful information on channel characteristics at the beginning of an exchange. Instead of forbidding the use of *bonjour,* the *Office* could help communication by encouraging its use and by suggesting that it should be spoken *before* the name of the organization rather than after it, since it is the most predictable word possible.

Speech output systems sometimes seem to have been designed as though they were to generate written rather than spoken messages. An example is provided by a system intended to allow passengers with a local bus company to find out by telephone when the next bus is due. Each stop has a unique number that corresponds to a telephone number. Dialing that number causes a computer-controlled speech output system to generate a message concerning the expected arrival of buses at that stop. On dialing a bus-stop number a passenger hears a message of the form:

"[Bus company] schedule for stop 8342. Route 3 in 5 and 25 minutes. Route 57 in 13 and 38 minutes. Thank you."

We have seen that more words are needed in spoken messages than in equivalent written messages if the speech is to be easily understood, but this style is even more terse than that of normal text. It is the style used in telegrams and telexes, where every extra word adds significantly to the cost. When spoken, it is hard to follow.

Since this is telephone speech, and since it is in addition peculiar, machine-generated speech, it is particularly important to give the listener a chance at the beginning of the message to adjust to the voice and to the channel. The first sentence in this message, even though it contains little useful information, is not predictable enough to allow the adaptation to take place. New users, unfamiliar with the stop number, will hear an apparently random sequence of digits, which is as unpredictable as anything that occurs in speech.

A more comprehensible message might be:

"Good morning. The next buses on Route 3 are due in 3 minutes and in 25 minutes; and on Route 57, they're due in 13 minutes and in 38 minutes. This information is for stop number 8342. Thank you, goodbye."

This alternative version contains about twice as many words as the original, so it might seem to be open to the objection that each enquiry would take twice as long. In fact, the stilted style of the original message forces a slow delivery and consequently makes the durations of the two messages comparable.

The same bus timetable enquiry system also provides an example of the pernicious effects at the acoustic level of projecting properties of speech onto text. It seems that words that might vary from message to message (destinations, times, etc) were recorded in isolation and are then concatenated to form the message. Worse still, syllables such as *teen* that are common to several words were recorded in isolation. If words were like text, this would of course be a reasonable thing to do, but we have seen that words are affected by their context and by their function in the sentence. The *3* in *8342* does not sound like the *3* in *Route 3* in normal speech. The effect of recording the words in isolation is to destroy the prosodic cues to sentence structure and to give each word prosodic cues corresponding to strong emphasis. It also affects the phonetic content to some extent. For example, the word *and* when used in fluent speech has a centralized vowel or often no vowel at all, something more like *'nd* or even *'n*. The kind of *and* produced in isolation is rare in fluent speech: it occurs only when the speaker wants to stress that important additional information is being added, as in:

"Sudso gets your dishes clean *and* it's kind to your hands"

It might seem that the emphasized version of the word would be easier to recognize in any context, since it contains clearer phonetic information. But this information is misleading when the listener is trying to understand the structure of the whole sentence. The centralization or deletion of the vowel in a normal *and* provides the useful information that this occurrence of the word does not need emphasis.

As we noted in the previous section, in a message made from concatenated isolated words any single word is perfectly clear but the message as a whole is not. In the case of the bus schedule system, even if users can manage to identify every word despite the confusing prosodic information, they generally find it difficult to retain the information about arrival times they were seeking.

In summary, the style used in spontaneous speech is different from that used in writing, and the differences do not arise solely from speech being less well planned than text. They reflect the different characteristics of the two modes of communication. Features present in speech but absent in text are often ignored.

SUMMARY

This chapter has tried to argue that speech is a special kind of signal. If we are to approach speech recognition, speech coding and speech synthesis effectively, we need to understand as much as we can about how we humans generate it and how we perceive it. In doing so, we have to avoid being misled by our internal impression of speech, which is not based on the raw signal but is the result of much sophisticated processing. We must above all avoid projecting properties of text onto speech.

ACKNOWLEDGMENTS

A few passages in this chapter have been published previously while the author was employed at the National Aeronautical Establishment, National Research Council, Canada, and funded by the Canadian Department of National Defence, DCIEM. Canadian Crown copyright exists on the publications containing these passages, and permission to reproduce them is gratefully acknowledged.

REFERENCES

1. HUNT M.J., "Studies of Glottal Excitation using Inverse Filtering and an Electroglottograph," *Proc. XI'th Intl. Congress of Phonetic Sciences,* Tallinn, Estonia, August 1-7, 1987, Vol 3, pp. 23-26.

2. MARKEL J.D.& GRAY A.H. *Linear Prediction of Speech,* Springer-Verlag, Berlin, 1976.

3. KLATT D.H., "Prediction of perceived phonetic distance from critical-band spectra : a first step" *Proc. IEEE Int. Conf. Acoust., Speech, Signal Processing,* Paris, May 1982, pp.1278-1281.

4. BORDEN, G.J & HARRIS, K.S., *Speech Science Primer* (2nd ed.), Williams & Wilkins, Baltimore, 1984.

5. LIBERMAN A.M, COOPER F.S, HARRIS K.S. & MACNEILAGE P.F "A motor theory of speech perception," *Proc. Stockholm Speech Comm. Seminar*, R.I.T., Stockholm, September 1962.

6. McGURK H. & MacDONALD J. "Hearing lips and seeing voices," *Nature* Vol. 264 #5588, pp.746-748, 1976.

7. LIEBERMAN, P. "Some effects of semantic and grammatical context on the production and perception of speech," *Language and Speech,* Vol. 6, 1963, pp.172-187.

8. HUNNICUTT, S. "Intelligibility versus redundancy — conditions of dependency," *Language and Speech,* Vol. 28, 1985, pp.47-56.

9. STUBBS, M. *Discourse Analysis: The Sociolinguistic Analysis of Natural Language,* Chicago, University of Chicago Press, 1983.

10. CHAPANIS, A. "Interactive Human Communication," *Scientific American,* Vol. 232, No. 3, March 1975, pp.36-49.

11. MARTIN, H. & PELLETIER, C. *Vocabulaire de la téléphonie,* Quebec City, Government of Quebec, June 1984, p.15.

CHAPTER 3

SPEECH CODING

Allen Gersho

Center for Information Processing Research
Dept. of Electrical & Computer Engineering
University of California
Santa Barbara, CA 93106

INTRODUCTION

Speech is the communication mechanism that distinguishes humans from lower animal forms and is an essential part of what allows man to function in civilization - our sophisticated ability to use language and communicate directly with one another via an acoustic channel. With the invention of the telephone by A. G. Bell, a major advance in human communication took place. Now we can communicate "in real-time" (not by writing letters or sending telegrams) with one another while geographically separated, perhaps around the world or in an aircraft or space vehicle. Of course the telephone was until recently based on analog communication: a simple modulation of an electric current in proportion to the instantaneous intensity of an acoustic signal. In recent decades digital communications emerged as a revolutionary new technology for the transportation of information and allowed us to develop new digital highways and superhighways carrying a variety of traffic such as data, video, and multiple channels of voice with greater reliability, cost effectiveness, privacy and security. Advances in error control and modulation techniques, including spread-spectrum and trellis-coded modulation allow reliable digital communication over radio channels that often suffer from interference, fading, and other degradations.

With the advent of rapidly increasing digital signal processing technology, it has recently become cost effective to use rather sophisticated speech coding algorithms in numerous commercial, government, and military communications applications. Speech coding is already being or becoming widely used in many storage applications where the communication process is not necessarily to transport voice from one geographical location to another but from one point in time to a later point

This work was supported in part by the National Science Foundation under grant NCR 8914741 and by Bell Communications Research, Inc., Bell-Northern Research, Inc., Rockwell International Corp., and the State of California MICRO program.

74

in time.

In this paper, we first describe some of the current and emerging applications of speech coding. Then we lead into the description of the main algorithms of interest today by starting with the basic ideas of predictive quantization, DPCM, LPC vocoders, and APC coders. Next, we introduce the idea of vector quantization, then come to excitation coding and coders based on analysis-by-synthesis coding and focus particularly on CELP or VXC type coders. Some recent developments of importance, vector sum excitation codebooks, low-delay VXC, and adaptive postfiltering are reviewed. Following this we introduce the use of phonetic segmentation in speech coding, a new approach that may contribute to the next generation of speech coders.

APPLICATIONS

Applications of speech coding today have become very numerous. A few examples are listed here: mobile satellite communications, cellular mobile radio, voice/data multiplexers for public and private networks, rural telephone radio carrier systems, audio for videophones or video teleconferencing systems, voice messaging networks, universal cordless telephones, audio/graphics conferencing, DCME digital circuit multiplexing equipment, voice memo wrist watch, voice logging recorders, and interactive PC software. New applications continue to emerge as digital signal processing technology makes very efficient compression increasingly cost effective.

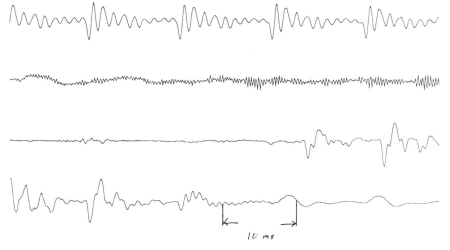

Fig. 1. Examples of Speech Waveforms

BASICS OF SPEECH CODING

The signals shown in Fig. 1 illustrate the great variety in the character of speech waveforms. Sometimes periodic or almost periodic, other times a mixture of periodic and random-like signals and sometimes the waveform appears like random noise. Shown in the figure is a 10 ms time interval. A speech coder operating, for example, at 4 kb/s must be able to describe any such 10 ms segment (80 samples) using only 40 binary digits in such a way that the segment will be reproduced with an accuracy sufficient to insure that it will sound very close to the original. Unlike PCM where 8 bits are used to code each sample, in such a low bit rate coder we have only 1/2 of a bit available per sample to describe the sound or the waveform. Of course there is no way to adequately describe the amplitude of a sample, even if an entire bit were available per sample (as in the case of an 8 kb/s coder). Thus we must use clever techniques to exploit redundancy across samples by introducing memory in the encoding process, so that we don't merely examine one sample at a time and code that sample (as in PCM), but we store up past samples, and/or information obtained from past samples to send out essential digital information that will help us to specify the current sample.

Speech coders have been traditionally grouped into vocoders (from "voice coders") and waveform coders. Today this dichotomy has become blurred with the current generation of so-called hybrid coders which embody some aspects of both of the above categories. Hybrid coders do attempt to reproduce the waveform, to some degree, while also describing key parameters that help to reproduce (synthesize) a sound perceptually similar to the original.

We assume the reader is familiar with PCM which, as used in telephony today, samples voice at 8,000 samples/s and codes each sample with an 8 bit word using a nonuniform quantizer based roughly on a logarithmic companding characteristic. Nothing further will be mentioned about this. Suffice it to note that a quantizer can be viewed as the cascade of an encoder (A/D converter) and a decoder (D/A converter). The encoder generates an index as a binary word specifying the amplitude level of the quantized value which approximates the input amplitude. Often the quantizer is viewed as a black box that generates both the index and the quantized level. The decoder (D/A) sometimes is called an 'inverse' quantizer and it simply maps the index into the reproduced level.

PREDICTIVE QUANTIZATION

A major advance in waveform coding of speech was the introduction of predictive quantization. Fig 2 shows the basic idea of this scheme. First, note that a quantity $\hat{X}_k$ is subtracted from the the input sample X_k forming a difference sample d_k. This difference is quantized and then the quantity $\hat{X}_k$ is added back to the quantized approximation of the difference sample d_k, producing a final output $\bar{X}_k$. Without giving any attention to how $\hat{X}_k$ is generated, it is evident that the error in

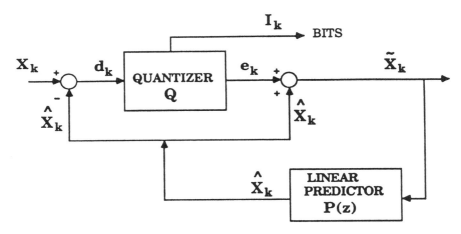

Fig. 2 Predictive Quantization

approximating the input sample X_k by $\tilde{X}_k$ is exactly equal to the error incurred by the quantizer in approximating the difference signal. This means that if we can somehow make $\hat{X}_k$ very close to X_k, the difference signal will be small, and fewer bits will be needed for quantizing d_k so as to make the overall error in approximating X_k by $\tilde{X}_k$ also small. The quantity $\hat{X}_k$ is chosen to be a linear prediction of X_k based on previously reproduced samples. The predictor has transfer function

$$P(z) = \sum_{i=1}^{n} a_i z^{-i}$$

The difference between the input sample and its predicted value, (based on the past information known to the decoder), is quantized and the index specifying the quantized level of this difference is sent to the decoder.

Note that the encoder implicitly contains a copy of the decoder. The actual decoder has an inverse quantizer which reproduces the sequence of quantized difference samples and feeds it into a filter with transfer function $[1-P(z)]^{-1}$, called the synthesis filter, to reproduce samples of the original signal X_k.

The performance gain of this structure is due to the prediction gain of the predictor, i.e. the ratio of variances of d_k to variance of X_k, or the factor by which the power of the input signal is reduced after removing the predictable error. This prediction gain in dB is what determines the performance improvement over straight PCM.

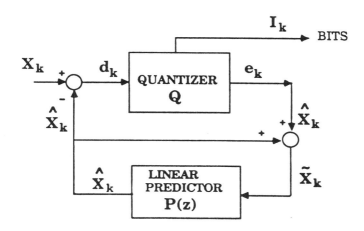

Fig. 3 Differential PCM in More Conventional Form

Fig. 3 shows the block diagram of a DPCM coder in a more conventional form, which is exactly the same scheme as in Fig. 2, only drawn in a less insightful way. By comparing the two figures, it is easily verified that they represent identical coders.

It is interesting to note that the DPCM decoder which generates speech from a sequence of difference samples models, in a primitive sense, the speech production mechanism in humans. The synthesis filter can be viewed as a model of the human vocal tract and the difference signal as a model of the acoustic excitation signal produced at the glottis. If the order of the predictor polynomial is reasonably high, (8 or higher) the synthesis filter indeed has a frequency response that reasonably corresponds to the overall filtering characteristics of the human vocal tract with its distinct spectral peaks, known as *formants*.

Of course, the human vocal tract is in constant movement and thus its frequency response varies substantially in time, from one phonetic sound unit, or phoneme, to another. Only over a time interval of a few milliseconds is it likely to be more or less constant. In Adaptive DPCM (ADPCM), the predictor is also time varying and thereby has a greater ability to model the speech production mechanism.

Another improvement in DPCM is the use of pole-zero prediction. Fig. 4 shows the predictive quantization structure modified by the use of two predictors $P_1(z)$ and $P_2(z)$. Each takes a linear combination of past values from its input. The new predictor, P_2, is applied directly to the quantized difference samples, while P_1 combines these with the preceding value of $\hat{X}_k$, to produce the current value of $\hat{X}_k$. Note that the corresponding decoder structure, also shown in Fig 4, has a pole-zero synthesis filter, where P_2 contributes zeros and P_1 poles to the synthesis filter.

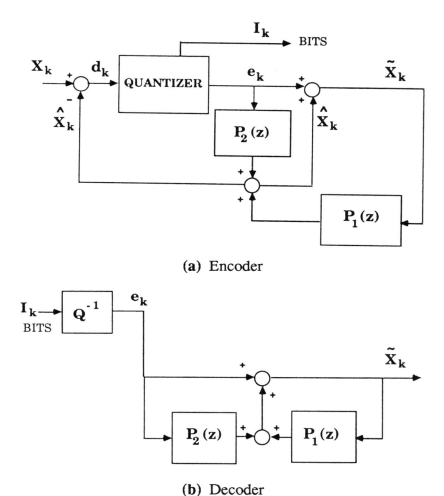

(a) Encoder

(b) Decoder

Fig. 4 Predictive Quantization (DPCM) with Pole-Zero Prediction

Indeed, the pole-zero filter may also provide a more versatile model of the human vocal tract if indeed a suitable number of poles and zeros were used and if the synthesis filter is adaptive (and thus time varying) to track the changing shape of the vocal tract. The CCITT 32 kb/s ADPCM standard, based on this structure, has 6 zeros and 2 poles and performs backward adaptation to make the two predictors track the time-varying statistics of the speech.

LPC VOCODER

In an entirely different approach to speech coding, known as parameter-coding, analysis/synthesis coding, or vocoding, no attempt is made at reproducing the exact speech waveform at the receiver, only a signal perceptually equivalent to it. Early versions of this approach included formant synthesizers and so-called "terminal analog synthesizers". However, the most widely used form today was partly motivated by recognizing the DPCM decoder as a model of the speech production mechanism. The idea is to replace the quantized difference signal by a simple excitation signal which at least crudely mimics typical excitation signals generated in the human glottis.

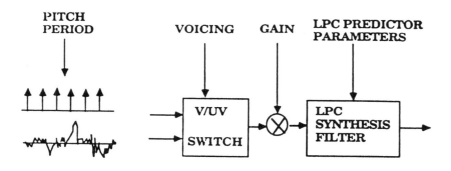

Fig. 5 LPC Vocoder - Decoder

Figure 5 illustrates the decoder structure of an LPC Vocoder. (LPC stands for Linear Predictive Coding.) The encoder sends a very modest number of bits to the decoder to describe each successive frame of the speech to be synthesized. A frame is a time segment typically 20 to 25 ms long. The excitation is specified by a one bit voicing parameter which indicates whether the frame of speech is judged to be periodic or aperiodic. Periodic segments correspond to so-called *voiced* speech where the glottis periodically opens and closes producing a fairly regular train of *pitch* pulses to the vocal tract. If the frame is voiced, the encoder also sends an estimate of the *pitch period* which typically ranges from 3 to 18 ms. The decoder locally generates one of two excitations, a periodic train of impulses at the pitch period, or (for unvoiced frames) a random noise excitation signal. A gain value must also be transmitted to specify the correct energy level of the current frame. Thus the set of parameters specified for the synthesis filter in each frames are: voicing decision, pitch (if appropriate), LPC coefficients (typically 10) and gain. The encoder of an LPC vocoder, also shown in Fig. 6, performs computations on each frame of input speech to determine the set of parameters needed by the decoder.

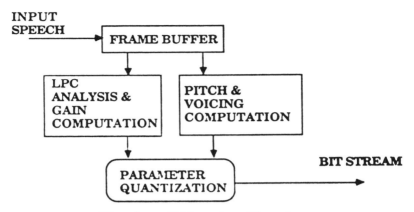

Fig. 6 LPC Vocoder - Encoder

The linear predictor described here and in the context of DPCM is often called a *short-term* predictor or *formant* predictor. For later convenience we denote the short-term predictor by $P_s(z)$ where s indicates short. These names illustrate the fact that the predictor exploits the short-term correlation in nearby samples of the speech waveform, and the fact that it is the short-term correlation which characterizes the formants dominating the envelope of the speech spectrum. Generally three or four principal formants are evident in examining the Fourier transform of a speech frame. The formant synthesis filter has a frequency response whose magnitude closely corresponds to the envelope of the spectrum. The transfer function of this synthesis filter is $[1 - P_s(z)]^{-1}$.

Note that the vocoder scheme does not actually attempt to encode the speech waveform but only extracts some parameters or features that partially characterize each frame. Thus it does not have the ability to reproduce an approximation to the original waveform. Nevertheless, it can synthesize clear, intelligible speech at the very low bit-rate of 2400 b/s. Such vocoders have served for years as the underlying technology for secure voice terminals, which include the functions of encrypting a bit stream and digital modulation into an analog voiceband signal suitable for transmission over an analog telephone connection.

PITCH PREDICTION

Another fundamental technique that has had a major impact on speech coding is the use of long-term or "pitch" prediction. The periodic, or nearly-periodic character of a speech segment suggests that there is considerable redundancy that can be exploited by predicting current samples from samples observed one period earlier. Because this periodicity is closely associated with the so-called fundamental frequency or *pitch* of voiced speech, the number of glottal openings per second, the repetition period is often called the *pitch period*. A long-term predictor or "pitch predictor" can be directly used to remove the periodicity when the period is known. The phrase "long-term" refers to the relatively large delay (many samples) used in pitch prediction compared to the small values for the short-term predictor. Thus, a pitch predictor typically has the transfer function

$$P_L(z) = \sum_{j=-i}^{i} \alpha_i z^{-m-i}$$

where m is the pitch period measured in samples, i is a small integer, and α_i are coefficients. Often a single tap predictor is used so that $i = 0$. The filter structure with transfer function $1 - P_L(z)$ removes periodicity, and thereby redundancy, by subtracting the predicted value from the current sample. This gives rise to a pitch synthesis filter, with the inverse transfer function $[1-P_L(z)]^{-1}$ which introduces a periodic character to an aperiodic input. We shall see how the pitch synthesis filter or *long-term* synthesis filter will play an important role in the new generation of speech coders.

The computation of the pitch predictor parameters, i.e. the pitch period and predictor coefficients, can be performed by the encoder in a manner similar to that used for LPC analysis where the buffered input speech is used to compute the predictor parameters. This is called an *open-loop* pitch analysis in contrast with a more recent method, to be described later, which optimizes the pitch predictor by directly measuring its impact on the overall quality of the speech reproduced by the decoder.

ADAPTIVE PREDICTIVE CODING (APC)

The oldest waveform coding technique which makes use of pitch prediction can be viewed as a sophisticated version of ADPCM. One version of an APC encoder is shown in Fig. 7. It clearly resembles the predictive coder of Fig. 2. In fact, the main difference in this structure is the addition of a pitch predictor to further remove redundancy from the input samples prior to quantization. In this scheme, we subtract from the input sample a short term prediction $\overset{\ast}{X}_k$ and then subtract a long-term prediction $\hat{X}_k$ to produce a difference signal d_k that has very little redundancy compared to the original sequence of speech samples X_k. Note that with this structure, the exact same prediction values are added back to the quantized difference signal e_k so that we have, as in DPCM, the property that the overall error between the original speech and the reconstructed speech $\tilde{X}_k$ is equal to the quantization error $d_k - e_k$.

A crucial distinction between APC and DPCM, not indicated in the figure, is that the short- and long-term predictors are updated for every frame, by directly computing the necessary parameters from a frame of speech stored in an input buffer prior to being encoded. This implies that side information describing the predictor parameters must be multiplexed with the bits produced by the quantizer to specify

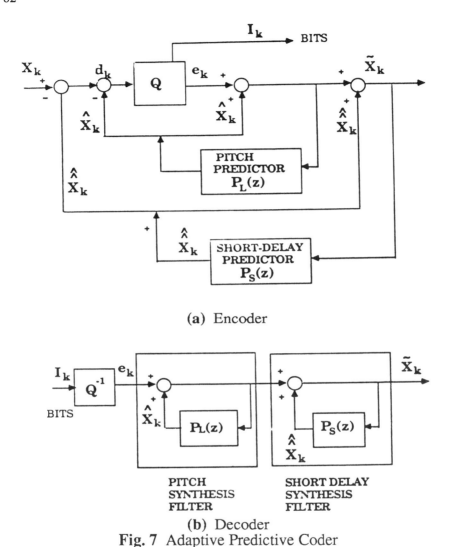

(a) Encoder

(b) Decoder

Fig. 7 Adaptive Predictive Coder

the difference signal often called the prediction *residual* signal. In fact, in typical APC coders a rather low bit-rate is found to be adequate to code the residual signal

The decoder for this APC scheme is also shown in Fig. 7 and it is evident that it reproduces the same sample sequence $\tilde{X}_k$ as generated in the encoder.

What is most noteworthy about the decoder structure is that the speech is being regenerated or *synthesized* by applying a signal e_k to a cascade of two synthesis filters. If a reasonably good job was done in determining prediction parameters and updating them at a reasonably frequent rate, e.g., a frame rate of 20 ms, it is found that this signal is very closely described as white Gaussian noise. Thus in effect, we are synthesizing speech from a time-varying speech production filter by applying to it a particular white noise excitation. This paradigm will recur again in subsequent discussions.

Various enhancements of APC have been developed, and in particular, quantization of the residual combined with entropy coding is often used. The APC structure can be modified by interchanging the role of long- and short-term prediction. APC speech coders have been implemented and used in the 1970s at typical bit rates of 9.6 kb/s and 16 kb/s. In the past decade, however, APC has gradually diminished in interest due to the emergence of newer and more powerful speech coding methods.

VECTOR QUANTIZATION

It has become recognized in the past decade that the efficient coding of a vector, an ordered set of signal samples or parameter values describing a signal, can be achieved by pre-storing a codebook of predesigned code vectors. For a given input vector, the encoder then simply identifies the address, or index, of the best matching code vector. Note that this is in essence a pattern matching algorithm. The index, as a binary word, is then transmitted and the decoder replicates the corresponding code vector by a table-lookup from a copy of the same codebook. In this way, the vector components are not coded individually as in scalar quantization, but rather all at once. Considerable efficiency is achieved, fractional bit rates (bits per vector component) become possible, and the average distortion (i.e., average squared error per component) for a given bit rate gets much reduced. Fig. 8 illustrates the basic idea of vector quantization (VQ).

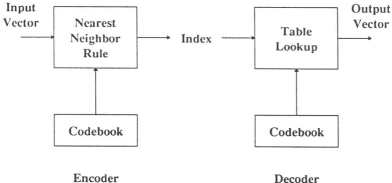

Fig. 8 Vector Quantization

The first major application of VQ to speech coding was reported by [1] where the bit rate of an LPC vocoder was substantially reduced by applying VQ to the LPC parameters. Subsequently VQ found its way into waveform coding as well and in particular a generalization of DPCM using vector prediction together with VQ was reported in [2]. Today VQ is a well-established and widely used technique. It has been applied to the efficient coding of the LPC parameter set, the pitch predictor filter parameters, as well as to Vector PCM (VPCM), the coding of a waveform by partitioning it into consecutive blocks (vectors) of samples. For a review of the use of VQ in speech coding, see [3]. An extensive treatment of the theory and technique of VQ for signal compression is given in[4].

OPEN LOOP VECTOR PREDICTIVE CODING

To illustrate the use of VQ, let us return to the APC scheme described above and consider that the largest contribution to the bit-rate of an APC coder is the coding of the residual waveform. However, in the structure of Fig. 7 the residual is generated only one sample at a time, and the next residual sample depends on feeding back the previous sample for obtaining the next short-term prediction. Thus the structure is not immediately amenable to VQ which requires storing up a block of residual samples before performing the pattern matching operation. There are two ways to circumvent this obstacle. One is based on a vector generalization of ADPCM introduced in [2] and extended to a vector version of APC in [5]. The other is simply to modify the encoder structure by removing the feedback around the quantizer, and generate the prediction residual by an open-loop method as is shown in Fig. 9. Note that the decoder has the same synthesis filter structure as that of the more conventional APC scheme. Here VPCM is applied to the residual signal, and since many of its samples may be encoded by a few bits, fractional bit rates (i.e. less than 1 bit per sample) can be attained.

Although this scheme has been applied by several researchers to speech coding, it suffers from one major disadvantage. Unlike the previous APC scheme, the overall error between original and reproduced speech in this coder is not equal to the error produced by the quantizer. Ordinarily, a VQ codebook is optimally designed to minimize the average distortion between input and reproduced vectors and encoding is performed by simply selecting the code vector best matching given input vector. In this coding scheme this implies that the reproduced residual is made to approximate the unquantized residual as closely as possible. However, this is *not* an optimal strategy, since our objective is to make the reproduced *speech* as close as possible to the original speech. With the predictor filters time-varying, these turn out not to be identical criteria, as the relationship between the error in quantizing the residual to the error in reproducing the original speech is a very complex one and varies from frame to frame.

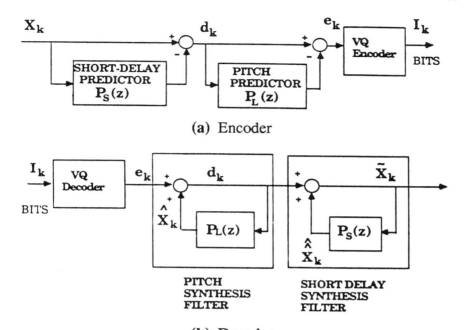

(a) Encoder

(b) Decoder

Fig. 9 Residual Encoding with Vector Quantization

These observations suggest that regardless of whether we use scalar or vector quantization or any other mechanism for digitally specifying an excitation signal for the decoder, the main task for the encoder is to figure out what excitation will do the best job of reproducing the original speech. The encoder structure of Fig. 9 incorporates a somewhat *ad hoc* mechanism for selecting an excitation vector from the codebook, which focuses narrowly on the residual signal, rather than on the speech itself. This is an intrinsic limitation of the open loop structure.

Let us therefore discard this encoder, and consider what is the best possible structure that can be used to supply data to the decoder given in Fig. 9. This perspective has led to a new generation of coding techniques, often called *hybrid* coding methods, which are based on the use of *analysis-by-synthesis* to determine the best excitation signal that will lead to an effective reproduction of the original speech.

ANALYSIS-BY-SYNTHESIS EXCITATION CODING

We now examine the most important family of speech coding algorithms known today, described as *Analysis-by-Synthesis Excitation Coding* or more concisely *Excitation Coding*. Consider the general decoder structure of Fig. 10, consisting of a synthesis filter (usually a cascade of both long- and short-term filters) to which is applied an excitation signal which is somehow specified by bits sent by the encoder.

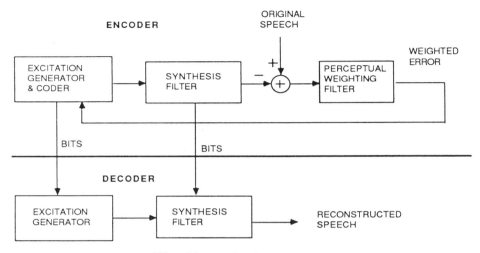

Fig. 10 Excitation Coding

The synthesis filters are periodically updated, usually by separate side information from the transmitter. The LPC analysis task is classical and straightforward, and we pay no further attention to it here. The open-loop method for computing the pitch predictor, which yields the synthesis filter parameters, was described earlier.

The encoder contains a copy of the decoder so that for any excitation waveform it can generate the same speech signal as the decoder would. Given a bit allocation and a mechanism for generating such waveforms, the encoder actually generates by trial and error all possible excitation signals for each time segment. The key idea here is that we try a large family of possible excitation segments and then apply each member in turn to the synthesis filter (the speech production model). For each synthesized segment we can compute a quantitative distortion measure, which indicates how badly the segment differs from the intended original. This process is repeated until the best excitation segment is found. Then, and only then, is the binary word specifying the best excitation segment transmitted.

The task of finding an appropriate excitation signal copying the decoder at the encoder, can be viewed as an *analysis* process, since in some sense we are extracting an appropriate excitation signal from the original speech. The method is called *analysis-by-synthesis* because this is done by synthesizing the speech segment that each candidate excitation would produce to examine how well it reproduces the original speech.

There are three principal mechanisms for generating excitation signals for this class of coding systems, known as tree or trellis coding, multipulse coding, and VQ. While all three are of interest, the third is most widely used, and we focus on this approach in the sequel. We shall refer to the generic coding algorithm for the use of a VQ codebook as Vector Excitation Coding (VXC); it is also known as Code-Excited Linear Prediction (CELP). This coding approach has led to many powerful speech coders for bit rates ranging from 4.8 to 16 kb/s.

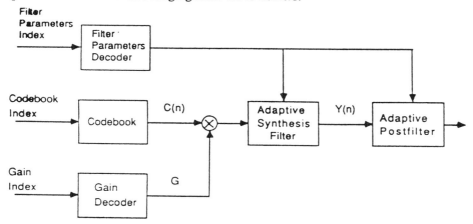

Fig. 11 VXC Decoder

VECTOR EXCITATION CODING

A generic VXC decoder structure is shown in Fig. 11. It is natural to describe the decoder first since it determines how the speech can be synthesized from transmitted data. Then encoder is in a sense a servant of the decoder, since its job is to examine the original speech and determine the best data to supply the decoder. The decoder receives and demultiplexes the data needed to vector, and in addition, a gain-scaling factor. A standard technique in VQ is to take advantage of the fact that owing to the wide dynamic range of speech, similarly shaped waveform portions may occur with different amplitudes, so that one may attribute to each segment a "gain" and a "shape" property. These attributes can then be handled separately via different codebooks, avoiding the ineffient duplication of waveform segments of similar shape, differing only in energy. By this method both codebook size and search complexity can be reduced.

It has been found empirically that the parameters of the synthesis filter need to be updated less frequently than new excitation vectors need to be supplied. For a 4.8 kb/s coder a typical frame size, i.e. the time span between successive updates of the synthesis filter, is 20-30 ms, while the excitation vector dimension, called *subframe*,

88

may be a quarter of this. For higher bit-rate coders there may be even more sub-frames in a frame.

For each subframe, the decoder receives a sequence of $c + g$ excitation code bits which identify a pair of indexes which specify one of 2^c excitation code vectors and one of 2^g gain levels, both by means of a table-lookup procedure. This leads to a gain-scaled excitation vector with dimension k. This vector is serialized as k successive samples and is applied to the synthesis filter. The filter is clocked for k samples, feeding out the next k samples of the synthesized speech; then it is "frozen" until the next scaled excitation vector is available as the next input segment to the synthesis filter.

In many applications an adaptive postfilter is added to the decoder as a final postprocessing stage, to enhance the quality of the recovered speech. This filter is adapted to correspond to the short term spectrum of the speech. We shall later describe the operation of the adaptive postfilter; however, for now we ignore it since it is not a fundamental or essential component of VXC.

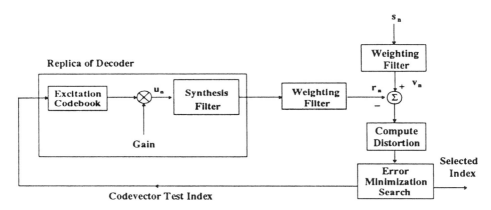

Fig. 12 VXC Encoder

The VXC Encoder

The VXC encoder structure is shown in Fig. 12. We describe its operation in the simplest way, while ignoring the many short cuts and tricks which greatly reduce the complexity involved in the search process. The encoder receives input speech samples which are grouped into blocks of k contiguous samples, each regarded as a vector. On the arrival of each such vector, the task of the encoder is to determine the next $c + g$ bits of data to be transmitted to the decoder so that the decoder will then be able to synthesize a reconstructed output speech vector that closely approximates the original input speech vector.

This implies that the encoder embody a replica of the decoder, as shown in Fig. 12, which can locally generate each of the $N = 2^{c+g}$ possible speech vector candidates that the decoder would produce for the same transmitted data values. However, the replica decoder does not include the postfilter used in the actual decoder.

In order to search for a reproduction that is closest, in a perceptually meaningful sense, to the original speech, a *perceptual weighting filter* is used to modify both the original input speech and the reconstructed output speech vector before the distortion between the two is measured. The weighting filter is combined with the synthesis filter to give a weighted synthesis filter with a modified transfer function that is distinct from the synthesis filter used The weighting filter is determined by the prediction error filter $Q(z) = [1 - P_s(z)]$ as computed in each frame. It has the transfer function $W(z) = Q(z)/Q(\beta z)$ where β is a parameter with typical value of 0.8. The effect of this filter is to enhance the observed quantization error at spectral valleys and reduce the error at spectral peaks during the search for the best excitation vector. As a result, the search is guided by a perceptually more meaningful indicator of distortion, so that the synthesized speech has somewhat more quantization noise at frequencies where the formant peaks occur and less where the valleys occur. This helps to "hide" some of the quantizing noise in regions where it is less audible due to masking effects of the human auditory system; at the same time it reduces the noise in spectral valleys where it would be more audible. Speech samples emerging from the weighting filter are also configured into corresponding vectors of k contiguous samples, called "weighted speech vectors"

Since the replica decoder is operating repeatedly in the search process, we must ensure that each candidate output speech vector, corresponding to a candidate data index pair being tested, is produced under the same conditions as will be present when the actual decoder generates the next output vector. After each test of a candidate index pair, the memory state of the replica decoder has changed and is no longer at the correct initial condition for the next test. Therefore, before generating each of these candidates, the memory in the replica decoder (including the perceptual weighting filter) must also be reset to the correct initial conditions.

The error minimization search module sequentially generates a pair of test indexes, corresponding to a particular pair of code vector and gain level. These are fed to the replica of the decoder which generates a synthesized speech vector that would be produced by the actual decoder if this index pair were actually transmitted. The replica decoder is initialized by setting the weighted synthesis filter memory to those initial conditions that were determined after the prior search process was completed. Then, the test index, is applied to the excitation codebook and the gain index to the gain codebook, yielding a gain and an excitation vector. The gain scaled excitation vector is then applied to the weighted synthesis filter to produce the output vector r_n. The vector r_n is then subtracted from the input speech vector v_n and the distortion between these two vectors, i.e., the sum of the squares of the components of the difference vector, is computed by the distortion computation module. This error value is applied to the search module which stores the distortion value, compares it with the lowest distortion value obtained so far in the current search process, and, if appropriate, updates the lowest distortion value and the corresponding vector index.

VECTOR SUM EXCITATION CODEBOOKS

A question of practical importance, is how the quality of a given VXC coder can be improved if more bits are made available and to which components of the coder should these bits be assigned for the maximum benefit. It is generally recognized, that the best performance gain comes from increasing the codebook size. However adding just one bit per code vector doubles the codebook size and the corresponding search complexity. Thus, computational constraints of the available signal processor quickly force one to limit the codebook size and lead to alternative designs where the vector dimension is reduced and more bits are given to synthesis filter parameters. The use of specially constrained codebook structures offers the possibility of larger codebooks and significant performance improvements while maintaining tolerable complexity.

Gerson and Jasiuk recently introduced technique for reducing the complexity of the excitation codebook search procedure[6]. Rather than have each of M code vectors be independently generated either randomly or by a design procedure, they design b *basis* vectors and then generate the $M = 2^b$ code vectors by taking binary linear combinations of the basis vectors. The resulting coding algorithm, a derivative of VXC, is called Vector Sum Excited Linear Prediction (VSELP) and an 8 kb/s version of this algorithm has been adopted as a standard for the U.S. cellular mobile telephone industry. We next explain the basic idea of this technique for fast codebook search.

Let v_i denote the i th basis vector of a given set of b basis vectors. The code vectors are then formed as

$$\mathbf{u}_\theta = \sum_{i=1}^{b} \theta_i \, \mathbf{v}_i$$

by taking all possible linear combinations where $\theta_i = \pm 1$ for each i. Thus each binary-valued vector θ determines a particular code vector $\mathbf{u}_\theta$. Naturally, the b bit binary word transmitted over the channel can simply correspond to a mapping of θ values with +1 being a binary 1 and −1 being a binary 0. Since the code vectors are so simply generated, b basis vectors need be stored rather than storing an entire codebook of M code vectors.

This special codebook structure can be searched very efficiently. Instead of finding the vector output of the weighted synthesis filter for each of the M codevectors, only the filtered output of the b basis vectors need be determined because from these any synthesized output can be readily obtained by addition. Furthermore the search for the optimal codevector becomes computationally simplified by noting that the mean-squared error between the weighted input vector and a filtered codevector depends in a simple manner on the values of θ_i. By ordering the b bit binary word in a Gray code, only one bit changes from one word to the next. This means that only a simple change is needed to compute the mean-squared error for the next candidate code vector from the previous candidate code vector.

The vector sum approach can be augmented by using multiple-stage VXC [7]. and joint optimization of the gains for each stage. The joint optimization becomes easy to implement with the vector sum codebooks [6] in conjunction with the use of orthogonalization. The idea of orthogonalization is that after one codebook is searched for the best codevector, c, the second codebook is replaced by subtracting from each of its codevectors a scaled version of c so that the resulting modified codevectors are all orthogonal to c. This then allows jointly optimal gain values for c and for the selected codevector for the second stage to be easily computed. The method is easily extended to more than two codebooks. See also[8].

CLOSED-LOOP PITCH SYNTHESIS FILTERING

An alternative and improved method of designing the long-term predictor (LTP) filter was first proposed for the multipulse excitation coder [9] and later applied to vector excitation coders [10] [11] [12]. proposed for multipulse excitation coding and subsequently applied to VXC. Although it is of higher complexity and requires a higher bit-rate, it does offer superior performance. Furthermore, when the closed-loop LTP is used, the size of the excitation codebook is reduced and hence the computational load is reduced.

The pitch lag and predictor coefficients of a closed-loop LTP are chosen in such way that the mean square of the perceptually weighted reconstruction error vector is minimized.

For a one-tap LTP, the predictor parameters can be determined two steps: (a) find the pitch lag m (from a predefined range) that maximizes a quantity that is independent of the prediction coefficient, and (b) compute the prediction coefficient from a simple formula.

In the closed-loop LTP method, the pitch lag ordinarily has to be greater or equal to the speech vector dimension in order to obtain the previous LTP output vector. Hence, the vector dimension, which is also the adaptation interval of the LTP, needs to be reasonably small to handle short pitch periods. Decreasing the adaptation interval increases the bit rate needed to code the LTP parameters.

Recently the use of *fractional* pitch prediction for speech coding was introduced[13]. The idea is to use interpolation to allow predictors which correspond to pitch lag values that are integer multiples of a fraction of the sampling interval. An increased bit allocation is needed to specify a fractional pitch value, however a notable benefit in performance is achieved in this way.

ADAPTIVE POSTFILTERING

As already discussed, the perceptual weighting filter is a valuable component of a VXC encoder since it exploits the masking effect in human hearing by removing quantization noise from exposed frequency regions where the signal energy is low, and "hiding" it under spectral peaks. At bit rates as low as 4.8 kb/s or less, however, the average noise level is quite high and thus it is not possible to simultaneously keep the noise below the masking threshold at spectral valleys as well as at formant frequencies. Since the formant peaks are more critical for perceptual quality, at low bit rates the weighting filter tends to protect these regions while tolerating noise above the threshold in the valleys. The technique of adaptive postfiltering attempts to rectify this by selectively attenuating the reproduced speech signal in the spectral valleys. This somewhat distorts the speech spectrum in the valleys but it also reduces the audible noise. Since a faithful reproduction of the spectral shape is perceptually much less important in the valleys than near formants, the overall effect is beneficial and leads to a notable improvement in subjective speech quality.

A more primitive form of adaptive postfiltering to enhance performance was applied to ADPCM by Ramamoorthy and Jayant[14] and to APC by Yatsuzuka [15]. Recently, an improved version for adaptive postfiltering was found [16] which is effective for VXC (or CELP).

For a filter to attenuate the spectral valleys, it must adapt to the time-varying spectrum of the speech. The synthesis filter parameters provide the needed information to identify the location of these valleys and are thus used to periodically update the postfilter parameters. since the LPC spectrum of a voiced sound typically has tilts downward at about 6 dB per octave, the corresponding all-pole postfilter will also have such a tilt causing undesirable muffling of the sound. This can be overcome by augmenting the postfilter with zeros at the same or similar angles as the poles but with smaller radii. The idea is to generate a numerator transfer function that compensates for the smoothed spectral shape of the denominator. The overall transfer function used for the postfilter in [16] is a pole-zero transfer function, given by:

$$H(z) = (1 - \mu z^{-1}) \frac{1 - P_s(z/\beta)}{1 - P_s(z/\alpha)}$$

Figure 13 shows the spectral magnitude of an all-pole filter $[1 - P_s(z/\alpha)]^{-1}$ for different values of α and for a particular LPC speech frame. Note the spectral tilt effect that arises here. For comparison, the frequency response of the pole zero postfilter is shown in Fig. 14 where the spectral tilt (and associated muffling effect) are substantially reduced. Since the transfer function of the postfilter changes with each speech frame, a time-varying gain is produced. To avoid this effect, an automatic gain control is used.

We can think of the reproduced speech coming into the postfilter as the sum of clean speech and quantizing noise. Although the postfilter is of course attenuating

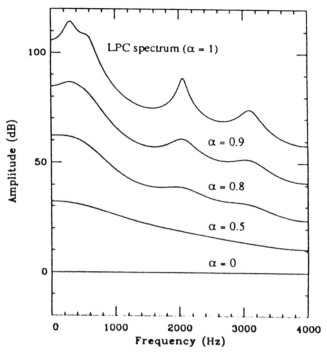

Fig. 13 Spectral Magnitude of All-Pole Postfilter $[1-P_s(z/\alpha)]^{-1}$ for different values of α and for a particular LPC speech frame.

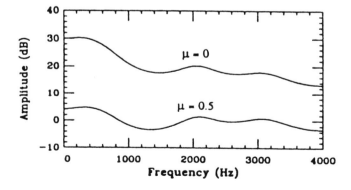

Fig. 14 Spectral Magnitude of Pole-Zero Postfilter

spectral valleys of both the speech and the noise, the distorting effect of the filter on the speech is negligible due to the low sensitivity of the ear to changes in the level of the spectral valleys. This has been verified by applying the original (uncoded) speech to the adaptive postfilter: the original and filtered speech sound essentially the same.

Though postfiltering clearly improves the performance of a single coder, when multiple stages of coding and decoding follow each other, the postfilter in each stage introduces a slight degradation that accumulates with the number of stages. postfiltering may thus not be desired for applications with tandeming.

Pole-zero adaptive postfiltering following the approach described above has been included in the U.S. digital cellular telephone standard for 8 kb/s speech coding and as an optional feature in the U.S. government standard for 4.8 kb/s speech coding. [17]. Both standards are derivatives of VXC.

LOW DELAY VXC

Vector Excitation Coding (VXC) combines techniques such as vector quantization, analysis-by-synthesis codebook searching, perceptual weighting, and linear predictive coding to successfully achieve good speech quality at low bit rates. However, one important aspect of coding has been ignored in the development of VXC or other conventional low bit rate excitation coding schemes; that is the *coding delay*. In fact, most existing speech coders with rates at or below 16 kbps require high delays in their operation, and cause various problems when they are applied to practical communication systems. In case of VXC, a large net coding delay, excluding computational delays, results from the use of buffering needed to perform the LPC and open-loop pitch analysis. Recently, new methods have been proposed to adapt synthesis filters without the high coding delay mentioned above while maintaining the quality of encoded speech.

With the conventional VXC scheme described above, the synthesis filter is adaptively updated every frame using what is sometimes called forward adaptation, the process of recomputing and updating
the desired filter parameters from the input speech. The use of the forward adaptation has two disadvantages: it requires transmission of side information to the receiver to specify the filter parameters and it leads to a large encoding delay of at least one analysis frame due to the buffering of input speech samples. The input buffering and other processing typically result in a one-way codec delay of 50 to 60 ms. In certain applications in the telecommunications network environment, coding delays as low as 2 ms per codec are required. In seeking new standards for speech coding, the CCITT adopted a performance objective of less than 2 ms for candidate 16 kb/s speech coding algorithms to be considered for a new standard intended to achieve the same quality as the 32 kb/s ADPCM standard, G.721. More recently, the CCITT initiated a study aimed at finding an 8 kb/s standard for speech coding with

essentially the same quality objectives as for 16 kb/s but allowing an algorithmic delay of up to 5 ms. Such low delays are not feasible with the established coders that are based on forward adaptive prediction coding systems. Although the 32 kb/s ADPCM algorithm, CCITT Recommendation G.721, satisfies the low delay requirement, it cannot give acceptable quality when the bit rate is reduced to 16 or 8 kb/s

An alternative solution is based on a recently proposed backward adaptation configuration. In a backward adaptive analysis-by-synthesis configuration, the parameters of the synthesis filter are not derived from the original speech signal, but computed by backward adaptation extracting information only from the sequence of transmitted codebook indices. Since both the encoder and decoder have access to the past reconstructed signal, side information is no longer needed for synthesis filters, and the low-delay requirement can be met with a suitable choice of vector dimension.

Two approaches to backward adaptation VXC have been studied, and they are classified as *block* and *recursive*. In the block algorithms, the reconstructed signal and the corresponding gain-scaled excitation vectors are divided into blocks (frames), and the optimum parameters of the adaptive filter are determined independently within each block. In the recursive algorithms, the parameters are updated incrementally after each successive excitation and reconstructed vector are generated.

To achieve the 2 ms delay objective, two versions of low were initially proposed to the CCITT. The first, called LD-CELP, is due to J. H. Chen. It uses a codebook of dimension 5 and a very high order block-adaptive short-term predictor computed by LPC analysis on the previously reproduced speech [18]. The second, called LD-VXC, has a codebook of dimension 4 and uses a recursive backward adaptation method for both a pole-zero predictor and for a pitch predictor [19]. With the standard sampling rate of 8 KHz used in telephony, a codebook of size 256 (8 bits) yields the rate of 16 kb/s . The second CCITT candidate was subsequently withdrawn and a modified version of the Chen algorithm is currently in the final stages of evaluation for the proposed 16 kb/s standard. Simulation results show that LD-VXC achieves an SNR of about 20 dB with either block or recursive adaptation. Transmission errors were also taken into account in the design of LD-VXC. With the help of leak factors and pseudo-gray coding, the performance of the coder only degrades slightly at 0.1% error rate, and intelligible speech is produced even at error rate as high as 1%.

For 8 kb/s low delay coding, modified LD-VXC and LD-CELP algorithms are reportedly under development at several research laboratories which are targeted for the CCITT 8 kb/s objectives. Two recent publications have already appeared describing alternative methods for low delay coding at 8 kb/s[13] and[13]. At the University of California in Santa Barbara, we have been developing a low-delay LD-VXC algorithm which gives a quality very close to that of the 8 kb/s VSELP algorithm (which is not a low delay algorithm). The lower bit rate led us to a vector dimension of 20 and the use of a multistage excitation technique[7] to obtain sufficient excitation quality without exorbitant complexity A forward adaptive closed-loop pitch predictor with either three tap or fractionally-spaced pitch prediction was found to be superior to backward pitch prediction. The algorithm uses a low-order block adaptive backward short-term predictor. In order to reduce the bit allocation for the pitch predictor, interframe predictive coding is employed. Efforts are continuing to further enhance the quality and robustness to transmission errors. The algorithm is described in [20] and similar ideas are reportedly under development at other laboratories. It is possible that an algorithm incorporating some of these methods may ultimately be adopted as an 8 kb/s standard by the CCITT.

VXC WITH PHONETIC SEGMENTATION

Although VXC achieves fairly high-quality speech at 4.8 kbps, the performance achieved with current VXC based algorithms degrades rapidly as the bit-rate is reduced below 4.8 kbps, leaving a substantial gap between the natural voice quality of VXC at 4.8 kbps and the synthetic quality attainable at 2.4 kbps (or higher) with an LPC vocoder. An important future direction for speech coding is to find coding algorithms that will achieve at 4 kb/s and below the natural quality attainable today with the best versions of VXC. One of the motivations for this interest is the next generation of digital cellular telephones where it is expected that a bit rate in the neighborhood of 4 kb/s will be required in order to meet the increasing channel capacity objectives.

One research direction that we have been studying, Phonetically-Segmented VXC (PS-VXC)[21], appears to show promise and might lead to a speech coder operating at bit rates significantly below 4.8 kb/s yet with a quality comparable to current 4.8 kb/s coders.

In this method, speech is segmented into a sequence of contiguous variable-length segments constrained to be an integer multiple of a fixed unit length. The segments are classified into one of six phonetic categories. This provides the front-end to a bank of VXC coders that are individually tailored to the different categories.

The motivation for this work derives from the fact that phonetically distinct speech segments require different coding treatments for preserving what we call *phonetic integrity*. With phonetic segmentation, we can assign the wide variety of possible speech segments into a small number of phonetically distinct groups. In each group, different analysis methods and coding strategies can be used to emphasize the critical parameters corresponding to important perceptual cues. It also becomes easier to identify each individual coding problem in isolated phonetic groups and optimize a multi-mode coding algorithm to suit various phonetic categories.

Table 1 summarizes the segment classification and coding structures used for these classes by specifying salient features and coder parameters for each of the six categories. Table 2 lists the bit-allocation for each category in PS-VXC. The details of the coding algorithm and recent improvements are reported in [21] and [22].

The three main segment types, if coded individually, would yield rates as follows: unvoiced — 3 kb/s, unvoiced/onset pairs — 3.6 kb/s, voiced — 3.6 kb/s. For typical speech files, the average rate is 3.4 kb/s, which could be achieved as a fixed rate with buffering of the encoder output. Alternatively, a fixed rate of 3.6 kb/s is readily attainable with some padding of the bit stream.

Informal listening tests indicate that the quality at a fixed 3.6 kb/s rate is roughly comparable to that of conventional VXC at 4.8 kb/s. Nevertheless, there is room for considerable improvement in both the coding algorithm for particular segment categories and in the definition and number of the phonetic classes used in the segmentation process. An end-to-end coding delay of approximately 100 ms (including overhead) is anticipated.

NONLINEAR PREDICTION OF SPEECH

Recently, a new method for the nonlinear prediction of one dimensional signals was introduced called Nonlinear Predictive Vector Quantization (NLPVQ). It involves the approximation of conditional expectations by using vector quantization and sample averaging. The main motivation for this research was the inherent limitation of linear prediction, which is the most commonly used technique in speech coding. Its simplicity and low computational complexity, as well as the availability of various adaptation techniques have made linear prediction an attractive method for speech data compression. However, linear prediction is known to be suboptimal for non-Gaussian signals such as speech, and does not make use of the higher-order statistics which may contain valuable information. Nonlinear prediction of speech in principle offers a way to enhance the redundancy-removing feature of linear prediction and provide a superior model of the speech production mechanism.

A novel method for the nonlinear prediction of speech was developed which does not require a parametric model of the predictor. When the constraint of linear estimation is abandoned, the best least-squares estimate of a random variable is its conditional expectation given the observed variables. Although in general this operation requires the knowledge of the joint pdf of the signal, VQ provides a way of approximating it using the training data directly. Instead of conditioning the desired random variable on past observables, it is conditioned on a quantized version of the past. As the quantizer resolution gets higher, the approximation will become more accurate. Hence, the estimate will approach, in an asymptotic sense, the optimal mean-square estimate. This is essentially a special case of Nonlinear Interpolative Vector Quantization (NIVQ) [23]. The observable vector is quantized to a codevector at the encoder. The encoder then generates an index which the decoder uses to address the predicted value of the next sample in a prediction table.

The method was tested for a known stationary signal with non-Gaussian statistics and nonlinear dependencies between adjacent samples. We found in this case that the prediction gain obtained with NLPVQ is remarkably close to the theoretical optimum. NLPVQ was then applied to voiced and unvoiced speech and tested for various prediction orders, codebook sizes, and training set sizes. In each case, the resulting prediction gains were compared with results obtained for linear predictors designed from the same training set. For unvoiced speech, the Segmental Prediction Gain (SEGPG) obtained with NLPVQ was found to be higher than that obtained by LP even for small codebook sizes. For voiced speech, the SEGPG of second-order NLPVQ exceeds that of the linear predictor at sufficiently large codebook sizes (12-14 bits).

The most dramatic result observed with this technique was the effect of NLPVQ on the residual signal. From informal listening tests, a major drop in the intelligibility of the prediction residual for NLPVQ was observed compared to LP. Moreover, the spectrum of the nonlinear prediction residual is significantly flatter than that of the LP residual. NLPVQ appears to be successful in eliminating the higher-order harmonics of the fundamental pitch frequency, thus potentially reducing the need for long term prediction.

While these results motivate further study of nonlinear prediction, practical application to speech coding may be premature. There are several nontrivial problems with nonlinear prediction that must be solved before it can become a useful technique for analysis-by-synthesis coding systems. Additional effort is necessary in the the joint optimization of the encoder and decoder, and the adaptation of the nonlinear predictor to time-varying signal statistics. Nevertheless, it is an interesting new direction in speech coding that is certainly worthy of further study.

CONCLUDING REMARKS

In this overview, we have only touched the surface of the rich and active field of speech coding. We have described some of the main concepts that underly speech coding algorithms of current interest today. In particular, linear prediction for both short and long term, analysis-by-synthesis, vector quantization, perceptual weighting for noise shaping, adaptive postfiltering, closed-loop pitch analysis, vector-sum codebook structures, and nonlinear prediction. No doubt in the next few years, there will be new advances that we cannot anticipate today.

The motivation for the continued activity in speech coding research is in large part due to the combination of two factors: the rapidly advancing technology of signal processor integrated circuits and the ever increasing demand for wireless mobile and portable voice communications. The technology permits increasingly complex and sophisticated signal processing algorithms to become implementable and cost effective. Mobile communications and the emerging personal communication network (PCN), with cordless portable personal telephones will increasingly stress the limited radio spectrum that is already pushing researchers to provide lower bit-rate and higher quality speech coding with lower power consumption, increasingly miniaturized technology, and lower cost. The insatiable need for humans to communicate with one another will continue to drive speech coding research for years to come.

References

[1]   A. Buzo, A. H. Gray, R. M. Gray, and J. D. Markel, "Speech Coding Based upon Vector Quantization," *IEEE Trans. Acoust., Speech, and Signal Processing*, vol. ASSP-28, no. 5, pp. 562-574, October 1980.

[2]   V. Cuperman and A. Gersho, "Vector Predictive Coding of Speech at 16 kbits/s," *IEEE Transactions on Communications*, vol. COM-33, pp. 685-696, July 1985.

[3]   A. Gersho, S. Wang, and K. Zeger, *Vector Quantization Techniques in Speech Coding,* Marcel Dekker, 1991.

[4]   A. Gersho and R. M. Gray, *Vector Quantization and Signal Compression,* Kluwer Academic Publishers, Norwell, Massachusetts, 1991.

[5]   J. H. Chen and A. Gersho, "Vector Adaptive Predictive Coding of Speech at 9.6 kb/s," *Proc. IEEE Inter. Conference on Acoust., Speech, and Signal Processing*, pp. 1693-1696, Tokyo, Japan, April 1986.

[6]   I. A. Gerson, M. A. Jasiuk, "Vector Sum Excited Linear Prediction," *IEEE Workshop on Speech Coding for Telecommunications*, Vancouver, September 1989.

[7]   G. Davidson, A. Gersho, "Speech Waveforms," *Proc. Inter. Conf. Acoust., Speech, & Signal Processing*, pp. 163-166, April 1988.

[8]   M. Johnson and T. Taniguchi, "Pitch-Orthogonal Code-Excited LPC," *Proc. IEEE Global Communications Conference*, pp. 542-546, Dec. 1990.

[9]   S. Singhal and B. S. Atal, "Improving Performance of Multi-Pulse LPC Coders at Low Rates," *Proc. IEEE Inter. Conf. Acoustics, Speech, and Signal Processing*, vol. 1, pp. 1.3.1-1.3.4, San Diego, March 1984.

[10] R. C. Ross and T. P. Barnwell, "The Self-Excited Vocoder," *Proceedings of IEEE International Conference on Acoustics, Speech, and Signal Processing*, vol. 1, pp. 453-456, Japan, April, 1986.

[11] P. Kabal, J.L. Moncet, and C.C. Chu, "Synthesis Filter Optimization and Coding: Applications to CELP," *Proc.IEEE Inter. Conf. Acoust., Speech, and Signal Processing*, vol. 1, pp. 147-150, New York City, April 1988.

[12] W. B. Kleijn, D. J. Krasinski, R. H. Ketchum, and Improved Speech Quality and Efficient Vector Quantization in SELP, *Proceedings of IEEE International*

Conference on Acoustics, Speech, and Signal Processing, vol. 1, pp. 155-158, New York, April, 1988.

[13] P. Kroon and B. S. Atal, and T. Moriya, L. G. Neumeyer, W. P. LeBlanc, and S. A. Mahmoud, "A Low-Delay 8 kb/s Backward-Adaptive CELP Coder," *Proc. International Mobile Satellite Conference, Ottawa%P 684-689*, vol. 2, pp. 15.16.1-15.16.4, Albuquerque, 1990.

[14] V. Ramamoorthy, N.S. Jayant, "Enhancement of ADPCM Speech by Adaptive Postfiltering," *Conf. Rec., IEEE Conf. on Commun.*, pp. 917-920, June 1985.

[15] Y. Yatsuzuka, S. Iizuka, T. Yamazaki, "A variable Rate Coding by APC with Maximum Likelihood Quantization from 4.8 bit/s to 16 kbit/s," *Proc. Inter. Conf. Acoust., Speech, & Signal Processing*, pp. 3071-3074, April 1986.

[16] J. H. Chen and A. Gersho, "Real-Time Vector APC Speech Coding at 4800 bps with Adaptive Postfiltering," *Proc. Int. Conf. on Acoust., Speech, Signal Processing Speech, and Signal Processing*, vol. 4, pp. 2185-2188, Dallas, April 1987.

[17] J.P. Campbell, Jr., V.C. Welch, T.E. Tremain, "An Expandable Error-Protected 4800 BPS CELP Coder (U.S. Federal Standard 4800 BPS Voice Coder)," *Proc. Inter. Conf. Acoust., Speech, & Signal Processing*, pp. 735-738, May 1989.

[18] J. H. Chen, "A Robust Low-Delay CELP Speech Coder at 16 kb/s," *Proc., IEEE Global Commun. Conf.*, November 1989.

[19] V. Cuperman, A. Gersho, R. Pettigrew, J. Shynk, J. Yao and J. H. Chen, "Backward Adaptive Configurations for Low-Delay Speech Coding," *Proc., IEEE Global Commun. Conf.*, November 1989.

[20] J.-H. Yao, J. J. Shynk, and A. Gersho, "Low Delay Vector Excitation Coding of Speech at 8 kbit/s," *Proc. IEEE Global Commun. Conf.*, submitted for publication, 1991. 1991

[21] Shihua Wang and Allen Gersho, "Phonetically-Based Vector Excitation Coding of Speech at 3.6 kbit/s," *Proc. IEEE Inter. Conf. Acoust., Speech, and Signal Processing*, Glasgow, May 1989.

[22] Shihua Wang and Allen Gersho, "Phonetic Segmentation for Low Rate Speech Coding," *Advances in Speech Coding*, Kluwer Academic Publishers, 1991..

[23] A. Gersho, "Optimal Nonlinear Interpolative Vector Quantization," *IEEE Trans. on Comm.*, vol. COM-38, No. 9, pp. 1285-1287, September 1990.

CHAPTER 4

VOICE INTERACTIVE INFORMATION SYSTEMS

J. L. Flanagan
CAIP Center, Rutgers University
New Brunswick, New Jersey

INTERACTIVE INFORMATION SYSTEMS

Society depends upon people being able to acquire, assimilate and react to information. As the complexities of society increase, so likewise have the technologies for providing information for decision making. Frequently these technologies are not well matched to human sensory capacities, and the user of such systems either is in danger of being inundated by vast quantities of information that are difficult to sift and digest, or of being burdened by complicated operating procedures for accessing the information. There is great incentive to make information systems easier and more natural to use and to provide systems with the intelligence for anticipating needs of a user.

Technologies and Sensory Capacities

Among the technologies supporting modern information systems is that of switched digital transmission. End-to-end digital connectivity is becoming globally pervasive and will permit new dimensions in communication, computation and information access. Transmission capacities span a broad range, from gigabits/sec rates for optical fiber, through hundreds of megabits/sec for electrical cable, to hundreds of kilobits/sec for cellular radio and basic rate ISDN. (See Fig. 1)

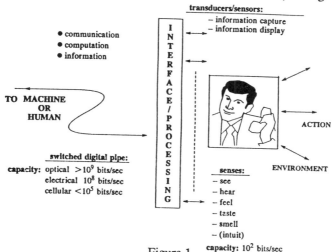

Figure 1
Information systems matched to human sensory capacities

These transports often bring information to a human user, whose sensory inputs depend primarily upon the modalities *of sight, sound* and *touch*. A significant issue is that the information capacity of these channels is several orders of magnitude smaller than the transport capacities delivering the information. (Experiments of different varieties estimate the processing rate at which the human can assimilate, perceive and react to these information inputs at less than 10^2 bits/sec.) An "impedance matching" function is needed to sense the needs of the moment and display in a "sensory-matched form" the necessary information. The technologies of automatic recognition and synthesis of speech, as well as image and tactile interaction can aid this process. We focus here primarily upon speech processing for the sound dimension.

NATURAL VOICE INTERFACES

Man's preferred means of communication is by natural voice. Giving interactive conversational ability to machines therefore opens opportunities for making technological complexities less of an obstacle for the human user. The understanding of automatic speech recognition and speech synthesis has now advanced to a point where machines can be given an 'ear' to listen and the intelligence to understand (in a sense) human-spoken commands, as well as a 'mouth' to speak and the intelligence to generate meaningful spoken responses. (See Fig. 2.)

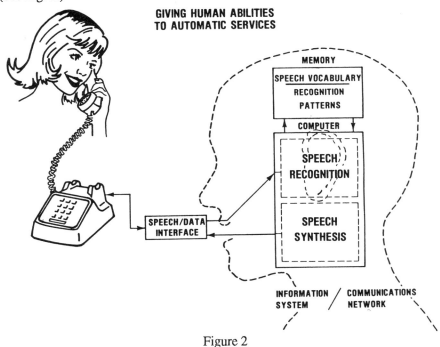

Figure 2
Human/machine interaction by voice

But, as yet, the capabilities of machines for intelligent interactive conversation are rigorously limited to application domains having well-defined specific tasks. In this instance the recognition vocabulary can usually be made tractably small and the grammar model pertaining to the task can be represented as a finite-state Markov process, which is, again, computationally tractable. For the voice answerback, text-to-speech synthesis has achieved the flexibility to serve well, with the synthetic output exhibiting high intelligibility, but typically lacking in natural quality. (See Fig. 3)

TASK-SPECIFIC VOICE CONTROL AND DIALOG

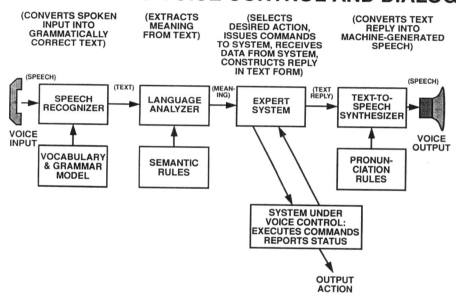

Figure 3
Generic task-specific system for conversational interaction

Speech Recognition

As shown in Fig. 3, the speech recognizer, with its stored vocabulary and grammar model, estimates the text string equivalent of the spoken input. The language model, with its set of semantic rules specifically tailored to a prescribed task, interprets the meaning of the command. This interpretation typically activates a control system and accesses the necessary data bases (which may constitute an expert system) to execute the request. Having carried out an action, the system generates an appropriate response in text form to report the status, and this reply is converted to spoken form by the text-to-speech synthesizer.

The speech recognition function to generate the text string estimate is essentially a pattern recognition operation. A well-established technique is matching the measured features of an unknown input utterance to a library of stored patterns corresponding to vocabulary items acceptable to the machine. (Fig. 4). Methods favored for this comparison include representing the time course of the speech spectrum by linear prediction coefficient (LPC) parameters, and using a dynamic programming method, or dynamic time warp (DTW), to adjust time scales for obtaining the closest match of the unknown input to the stored vocabulary patterns. More recently emerging from the laboratory is a statistical based technique for representing the vocabulary entries -- namely, the Hidden Markov Model (HMM), in which words or individual sounds are modeled by a sequence of states, their probabilities and the transition probabilities between states. The most probable sequence is computed and indicated by the recognizer.

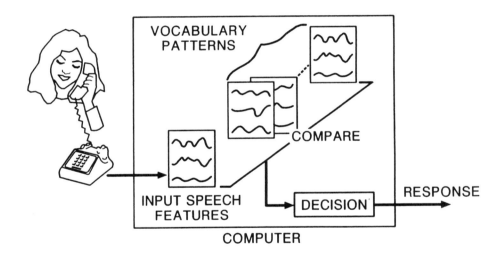

Figure 4
Pattern recognition approach to speech recognition

At present, the practical realities are that vocabularies of several hundred words can be reliably recognized in connected form when spoken by a wide variety of talkers (i.e., speaker independent performance). The machine grammars corresponding to specific tasks (such as making airline reservations, performing voice-activated computer calculations, and controlling hazardous material handling robots remotely by voice) typically are modelled as finite-state processes having several hundred states and arcs. Even with relatively small vocabularies (say, order of 100 words) the grammar can easily span billions of acceptable sentences. While the grammar is a restricted subset of natural language, it is usefully large for specific applications.

The semantic analyses corresponding to the same specific tasks typically embody several hundred rules to interpret the output word string from the recognizer and initiate actions.

Typical performance in the laboratory for isolated word and connected word systems is given in Fig. 5 a,b.

RECOGNITION PERFORMANCE
ISOLATED WORDS

VOCABULARY	ACCURACY (%)	
	SPEAKER DEPENDENT	SPEAKER INDEPENDENT
10 DIGITS	99	99
37 DIALER WORDS	99	—
39 ALPHA DIGITS	96	90
54 COMPUTER WORDS	—	96
129 AIRLINE TERMS	99	97
200 CITY NAMES	97	—
1109 WORDS-BASIC ENGLISH	96	—

RECOGNITION PERFORMANCE
CONNECTED WORDS
(SPEAKER INDEPENDENT SYSTEMS)

VOCABULARY	TASK	WORD ACCURACY (%)	STRING ACCURACY (%)	TASK ACCURACY (%) (WITH SYNTAX)
DIGITS (10)	7 DIGIT TELEPHONE NUMBERS	99	98	—
LETTERS (26)	DIRECTORY LISTING RETRIEVAL (17,000 NAMES)	80	17	90
AIRLINE TERMS (127)	AIRLINES TRAVEL INFORMATION	99	90	99

Figure 5a,b
Typical performance for laboratory speech recognizers

Computation requirements, as a very rough rule-of-thumb, vary by orders of magnitude in going from speaker-trained (or, speaker dependent) isolated-word systems, to connected word systems, to speaker independent systems. For example, for a 200-word vocabulary, the orders of computation are very roughly:

» isolated word, speaker dependent   2 mflops
» connected word, speaker dependent 20 mflops
» connected word, speaker independent 200 mflops

Large vocabulary systems (1,000 to 10,000 words) are now emerging from the laboratory, with expanded statistical models of language. While whole-word models remain presently the more robust technology (both for template and HMM methods), computational and storage requirements urge the use of smaller-than-word elements for the recognition process. Individual speech sounds are therefore modeled by HMM techniques in these developing systems. New knowledge in spoken language statistics - such as N-gram data - is simultaneously expanding the capabilities for language modeling, as well as opening new opportunities for automatic translating telephony.

Speech Synthesis

The voice response function for the interactive system is well served by text-to-speech synthesis, because of the flexibility in speaking virtually unrestricted text. The machine therefore is given great freedom in composing intelligent output messages - but, again, within the constraints of the task domain.

Speech synthesis from text is traditionally aided by a stored pronouncing dictionary and programmed letter to sound rules (Fig. 6). Stored dictionaries of the order of 30-50,000 entries are feasible, and separate dictionaries for proper names are also used. Generation of the sound sequence can be achieved from formant synthesis in which phonetic elements have associated target values, or from a separate store of fractional-syllable elements (about 2,000), which are excerpted from natural speech and LPC coded.

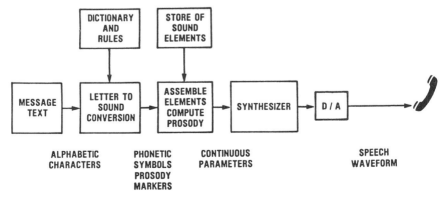

Figure 6
Elements of text-to-speech synthesis

The amount of computation and storage needed to support text synthesis is relatively modest, typically less than 10 mflops and several megabytes of memory. Some systems run on personal computers with adjunct hardware to perform the synthesizer function.

AUTODIRECTIVE MICROPHONE SYSTEMS

Voice interaction with an information terminal ideally should be as easy and as natural as face-to-face communication with a human. High-quality hands-free sound pick up, with no tether or encumbrance by hand-held or body-worn equipment, is therefore desirable. This capability becomes especially important if groups of users are assembled at a given terminal for conferencing purposes or collaborative activity.

Acoustic multipath (reverberation) in typical rooms, as well as ambient noise, are traditional obstacles to hands-free sound pick up. Adaptive noise filtering can be implemented to produce nulls of sensitivity in directions of noise sources, but phased-array beamforming of microphone arrays offers some distinct advantages in a reverberant enclosure. In this case the beam former "sees" fewer sound images in reflecting surfaces, and captures a signal with less reverberant interference. (see Fig. 7). With controllable delay hardware dedicated to each sensor in the array, the beam steering can be accomplished by modest compute power. Most of the computation is devoted to detecting the direction of a sound source, deciding whether it is speech or non-speech, and positioning the beam in the direction of the desired speech signal.

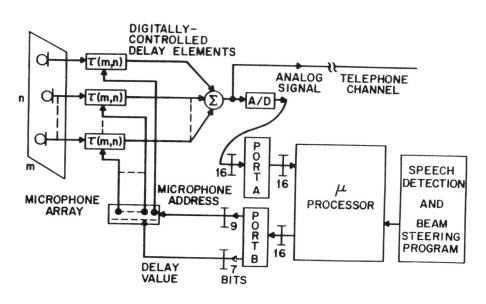

Figure 7
Autodirective beamforming for hands-free sound pick up

Depending upon the size of the room to be covered, and the number of participants, the phased array can be sized appropriately. Figure 8 shows a two-dimensional array of 408 beam-steered microphones. The high quality, low cost, and small size of electret transducers make such arrangement feasible, along with economical digital signal processors for effecting the control. This system actually forms two beams simultaneously for a 'track while scan' performance, and is used for sound pick up in a 300-seat auditorium.

Figure 8
Two dimensional phased-array 'track-while scan' beamformer for auditorium use

INTEGRATION OF VOICE IN MULTIMEDIA SYSTEMS

A number of information technologies are becoming advanced enough to support sophisticated features for interactive information systems. Prominent in these component technologies is speech interaction - made possible by speech recognition and synthesis. A challenge is how to integrate speech capabilities along with parallel advances in image processing and tactile interaction.

One recent prototype system for conferencing over switched digital telephone service is called **HuMaNet** (for Human Machine Network). It integrates hands-free autodirective sound pick up, speech recognition and synthesis for a limited task domain (conferencing over basic-rate ISDN), low bit-rate image coding and hypertext databases that are hierarchical, distributed, and networked. All features of the system are controlled by voice and are actuated by a connected-word recognizer receiving its speech input from a beam-steering microphone array. Small groups can set up conference calls over the two 64 kbps channels of basic rate ISDN, can request and display text and image information from local and remote data bases, and receive status reports by text synthesis from the system. For data bases that are restricted or privileged, a speaker verification procedure is incorporated.

All features are supported by 386 personal computers and a single file server. (See Fig. 9). Individual PC's are dedicated to specific functions, such as speech recognition and decoding of perceptually-compressed images. Stereo sound of 7KHz bandwidth is supported by fixed-steered line array microphones.

Figure 9
HuMaNet multimedia conferencing system for digital telephone use

A major issue in the utility of such capabilities is a human factors one. How should the technologies be combined in a complementary way, easy for humans to use? One measure of success is the way that the complexities of the system can be hidden from the user. Communication by natural spoken language is a major step in this direction.

PROJECTIONS FOR DIGITAL SPEECH PROCESSING

Explosive advances in the economy and capabilities of single-chip digital processors have underpinned virtually all practical applications in speech communication and human/machine interaction. These advances will continue over the next decade, as integrated circuits progress to fractional micron implementations. (See Fig. 10).

Processing Requirements
For Speech Applications

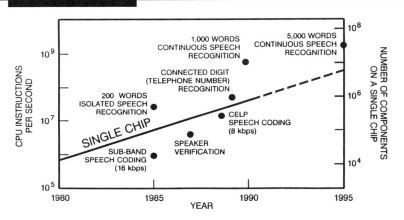

Figure 10
Projection of capabilities of single-chip digital signal processors

Today we have single-chip processors capable of 25-50 megaflops (at a cost of about one dollar per megaflop). The next decade is certain to see single-chip processors capable of a gigaflop, or more, hopefully at no increase over today's single-chip cost.

We conclude on this note, pointing out that a decade hence brings us exactly to 2001, and possibly to Arthur Clarke's vision of conversational machines such as HAL.

CHAPTER 5

SPEECH RECOGNITION BASED ON
PATTERN RECOGNITION APPROACHES

Lawrence R. Rabiner*

AT&T Bell Laboratories
Murray Hill, New Jersey 07974

Abstract. Algorithms for speech recognition can be characterized broadly as pattern recognition approaches and acoustic phonetic approaches. To date, the greatest degree of success in speech recognition has been obtained using pattern recognition paradigms. Thus, in this paper, we will be concerned primarily with showing how pattern recognition techniques have been applied to the problems of isolated word (or discrete utterance) recognition, connected word recognition, and continuous speech recognition. We will show that our understanding (and consequently the resulting recognizer performance) is best for the simplest recognition tasks and is considerably less well developed for large scale recognition systems.

INTRODUCTION

The ultimate goal of most research ~~is~~ for speech recognition is to develop a machine that had the ability to understand fluent, conversational speech, with unrestricted vocabulary, from essentially any talker. Although the promise of such a capable machine is as yet unfulfilled, the field of automatic speech recognition has made significant advances in the past decade [1-3]. This is due, in part, to the great advances made in VLSI technology, which have greatly lowered the cost and increased the capability of individual devices (e.g. processors, memory), and in part due to the theoretical advances in our understanding of how to apply powerful mathematical modelling techniques to the problems of speech recognition.

When setting out to define the problems associated with implementing a speech recognition system, one finds that there are a number of general issues that must be resolved before designing and building the system. One such issue is the size and complexity of the user vocabulary. Although useful recognition systems have been built with as few as two words (yes, no), there are at least four distinct ranges of vocabulary size of interest. Very small vocabularies (on the order of 10 words) are most useful for control tasks – e.g. all digit dialing of telephone numbers, repertory name dialing, access

---

*   Director, Information Principles Research, AT&T Bell Laboratories, Murray Hill, New Jersey 07974

control etc. Generally the vocabulary words are chosen to be highly distinctive words (i.e. of low complexity) to minimize potential confusions. The next range of vocabulary size is moderate vocabulary systems having on the order of 100 words. Typical applications include spoken computer languages, voice editors, information retrieval from databases, controlled access via spelling etc. For such applications, the vocabulary is generally fairly complex (i.e. not all pairs of words are highly distinctive), but word confusions are often resolved by the syntax of the specific task to which the recognizer is applied. The third vocabulary range of interest is the large vocabulary system with vocabulary sizes on the order of 1000 words. Vocabulary sizes this large are big enough to specify fairly comfortable subsets of English and hence are used for conversational types of applications – e.g. the IBM laser patent text, basic English, etc. [4, 5]. Such vocabularies are inherently very complex and rely heavily on task syntax to resolve recognition ambiguities between similar sounding words. Finally the last range of vocabulary size is the very large vocabulary system with 10,000 words or more. Such large vocabulary sizes are required for office dictation/word processing and language translation applications.

Although vocabulary size and complexity is of paramount importance in specifying a speech recognition system, several other issues can also greatly affect the performance of a speech recognizer. The system designer must decide if the system is to be speaker trained, or speaker independent; the format for talking must be specified (e.g. isolated inputs, connected inputs, continuous discourse); the amount and type of syntactic and semantic information must be specified; the speaking environment and transmission conditions must be considered; etc. The above set of issues, by no means exhaustive, gives some idea as to how complicated it can be to talk about speech recognition by machine.

There are two general approaches to speech recognition by machine, the statistical pattern recognition approach, and the acoustic-phonetic approach. The statistical pattern recognition approach is based on the philosophy that if the system has "seen the pattern, or something close enough to it, before, it can recognize it." Thus, a fundamental element of the statistical pattern recognition approach is pattern training. The units being trained, be they phrases, words, or subword units, are essentially irrelevant, so long as a good training set is available, and a good pattern recognition model is applied. On the other hand, the acoustic-phonetic approach to speech recognition has the philosophy that speech sounds have certain invariant (acoustic) properties, and that if one could only discover these invariant properties, continuous speech could be decoded in a sequential manner (perhaps with delays of several sounds). Thus, the basic techniques of the acoustic-phonetic approach to speech recognition are feature analysis (i.e. measurement of the invariants of sounds), segmentation of the feature contours into consistent groups of features, and labelling of the segmented features so as to detect words, sentences, etc.

To date, the greatest success in speech recognition have been achieved using the pattern recognition approach. Hence, for the remainder of this paper, we will restrict our attention to trying to explain how the model works, and how it has been applied to the problems of isolated word, connected word, and continuous speech recognition.

THE STATISTICAL PATTERN RECOGNITION MODEL

Figure 1 shows a block diagram of the pattern recognition model used for speech recognition. The input speech signal, $s(n)$, is analyzed (based on some parametric model) to give the test pattern, T, and then compared to a prestored set of reference patterns, $\{R_v\}$, $1 \leq v \leq V$ (corresponding to the V labelled patterns in the system) using a pattern classifier (i.e. a similarity procedure). The pattern similarity scores are then sent to a decision algorithm which, based upon the syntax and/or semantics of the task, chooses the best transcription of the input speech.

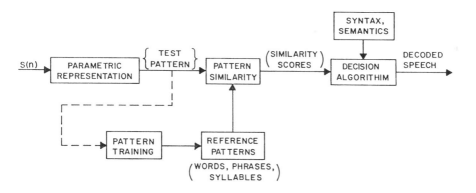

Figure 1. Pattern Recognition Model for Speech Recognition.

There are two types of reference patterns which can be used with the model of Fig. 1. The first type, called nonparametric reference patterns, are patterns created from one or more real world tokens of the actual pattern. The second type, called statistical reference models, are created as a statistical characterization (via a fixed type of model) of the behavior of a collection of real world tokens. Ordinary template approaches [6], are examples of the first type of reference patterns; hidden Markov models [7, 8] are examples of the second type of reference patterns.

The model of Fig. 1 has been used (either explicitly or implicitly) for almost all commercial and industrial speech recognition systems for the following reasons:

1. It is invariant to different speech vocabularies, users, feature sets, pattern similarity algorithms, and decision rules

2. It is easy to implement in either software or hardware

3. It works well in practice.

For all of these reasons we will concentrate on this model throughout this paper. In the remainder of this paper we will discuss the elements of the pattern recognition model and show how it has been used for isolated word, connected word, and for continuous speech recognition. Because of the tutorial nature of this paper we will minimize the use of mathematics in describing the various aspects of the signal processing. The interested reader is referred to the appropriate references [e.g. 6-15].

Parametric Representation

Parametric representation (or feature measurement, as it is often called) is basically a data reduction technique whereby a large number of data points (in this case samples of the speech waveform recorded at an appropriate sampling rate) are transformed into a smaller set of features which are equivalent in the sense that they faithfully describe the salient properties of the acoustic waveform. For speech signals, data reduction rates from 10 to 100 are generally practical.

For representing speech signals, a number of different feature sets have been proposed ranging from simple sets, such as energy and zero crossing rates (usually in selected frequency bands), to complex, complete representations, such as the short-time spectrum or a linear predictive coding (LPC) model. For recognition systems, the motivation for choosing one feature set over another is often complex and highly dependent on constraints imposed on the system (e.g. cost, speed, response time, computational complexity etc). Of course the ultimate criterion is overall system performance (i.e. accuracy with which the recognition task is performed). However, this criterion is also a complicated function of all system variables.

The two most popular parametric representations for speech recognition are the short-time spectrum analysis (or bank of filters) model, and the LPC model. The bank of filters model is illustrated in Fig. 2. The speech signal is passed through a bank of Q bandpass filters covering the speech band from 100 Hz to some upper cutoff frequency (typically between 3000 and 8000 Hz). The number of bandpass filters used varies from as few as 5 to as many as 32. The filters may or may not overlap in frequency. Typical filter spacings are linear until about 1000 Hz and logarithmic beyond 1000 Hz [9].

The output of each bandpass filter is generally passed through a nonlinearity (e.g. a square law detector or a full wave rectifier) and lowpass filtered (using a 20-30 Hz width filter) to give a signal which is proportional to the energy of the speech signal in the band. A logarithmic compressor is generally used to reduce the dynamic range of the intensity signal, and the compressed output is resampled (decimated) at a low rate (generally twice the lowpass filter cutoff) for efficiency of storage.

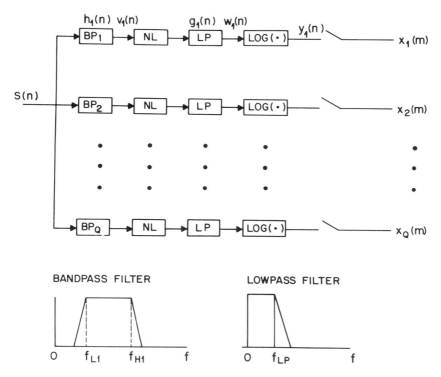

Figure 2. Bank of Filters Analysis Model.

The LPC feature model for recognition is shown in Fig. 3. Unlike the bank of filters model, this system is a block processing model in which a frame of N samples of speech is processed, and a vector of features is computed. The steps involved in obtaining the vector of LPC coefficients, for a given frame of N speech samples, are as follows:

1. preemphasis by a first order digital network in order to spectrally flatten the speech signal

2. frame windowing, i.e. multiplying the N speech samples within the frame by an N-point Hamming window, so as to minimize the endpoint effects of chopping an N-sample section out of the speech signal

3. autocorrelation analysis in which the windowed set of speech samples is autocorrelated to give a set of $(p + 1)$ coefficients, where p is the order of the desired LPC analysis (typically 8 to 12)

4. LPC analysis in which the vector of LPC coefficients is computed from the autocorrelation vector using a Levinson or a Durbin recursive method [10].

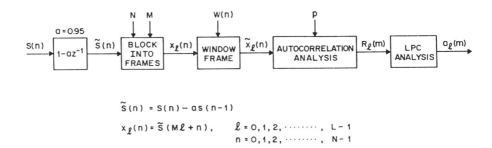

$$\tilde{s}(n) = s(n) - as(n-1)$$

$$x_\ell(n) = \tilde{s}(M\ell + n), \qquad \ell = 0, 1, 2, \cdots\cdots, L-1$$
$$n = 0, 1, 2, \cdots\cdots, N-1$$

Figure 3. LPC Analysis Model.

New speech frames are created by shifting the analysis window by M samples (typically $M < N$) and the above steps are repeated on the new frame until the entire speech signal has been analyzed.

The LPC feature model has been a popular speech representation because of its ease of implementation, and because the technique provides a robust, reliable, and accurate method for characterizing the spectral properties of the speech signal.

As seen from the above discussion, the output of the feature measurement procedure is basically a time-frequency pattern – i.e. a vector of spectral features is obtained periodically in time throughout the speech.

Pattern Training

Pattern training is the method by which representative sound patterns are converted into reference patterns for use by the pattern similarity algorithm. There are several ways in which pattern training can be performed, including:

1. casual training in which each individual training pattern is used directly to create either a non-parametric reference pattern or a statistical model. Casual training is the simplest, most direct method of creating reference patterns.

2. robust training in which several (i.e. two or more) versions of each vocabulary entry are used to create a single reference pattern or statistical model. Robust training gives statistical confidence to the reference patterns since multiple patterns are used in the training.

3. clustering training in which a large number of versions of each vocabulary entry are used to create one or more reference patterns or

statistical models. A statistical clustering analysis is used to determine which members of the multiple training patterns are similar, and hence are used to create a single reference pattern. Clustering training is generally used for creating speaker independent reference patterns, in which case the multiple training patterns of each vocabulary entry are derived from a large number of different talkers.

The final result of the pattern training algorithm is the set of reference patterns used in the recognition phase of the model of Fig. 1.

Pattern Similarity Algorithm

A key step in the recognition algorithm of Fig. 1 is the determination of similarity between the measured (unknown) test pattern, and each of the stored reference patterns. Because speaking rates vary greatly from repetition to repetition, pattern similarity determination involves both time alignment (registration) of patterns, and once properly aligned, distance computation along the alignment path.

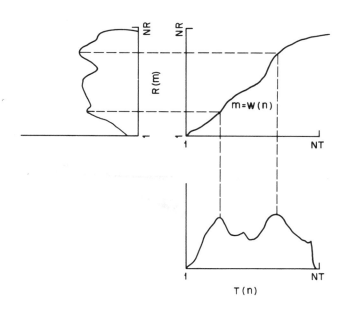

Figure 4. **Example of Time Registration of a Test and Reference Pattern.**

Figure 4 illustrates the problem involved in time aligning a test pattern, $T(n)$, $1 \leq n \leq NT$ (where each $T(n)$ is a spectral vector), and a reference pattern $R(m)$, $1 \le m \le NR$. Our goal is to find an alignment function, $m = w(n)$, which maps R onto the corresponding parts of T. The criterion for correspondence is that some measure of distance between the patterns be minimized by the mapping w. Defining a local distance measure, $d(n, m)$, as the spectral distance between vectors $T(n)$ and $R(m)$, then the task of the pattern similarity algorithm is to determine the optimum mapping, w, to minimize the total distance

$$D^* = \min_{w(n)} \sum_{i=1}^{NT} d(i, w(i)) \tag{1}$$

The solution to Eq. (1) can be obtained in an efficient manner using the techniques of dynamic programming. In particular a class of procedures called dynamic time warping (DTW) techniques, has evolved for solving Eq. (1) efficiently [6]. The above discussion has shown how to time align a pair of templates. In the case of aligning statistical models, an analogous procedure, based on the Viterbi algorithm, can be used [7,8,16].

Decision Algorithm

The last step in the statistical pattern recognition model of Fig. 1 is the decision which utilizes both the set of pattern similarity scores (distances) and the system knowledge, in terms of syntax and/or semantics, to decode the speech into the best possible transcription. The decision algorithm can (and generally does) incorporate some form of nearest neighbor rule to process the distance scores to increase confidence in the results provided by the pattern similarity procedure. The system syntax helps to choose among the candidates with the lowest distance score by eliminating candidates which don't satisfy the syntactic constraints of the task, or by deweighting extremely unlikely candidates. The decision algorithm can also have the capability of providing multiple decodings of the spoken string. This feature is especially useful in cases in which multiple candidates have indistinguishably different distance scores.

Summary

We have now outlined the basic signal processing steps in the pattern recognition approach to speech recognition. In the next sections we illustrate how this model has been applied to problems in isolated word, connected word, and continuous speech recognition.

RESULTS ON ISOLATED WORD RECOGNITION

Using the pattern recognition model of Fig. 1, with an 8th order LPC parametric representation, and using the non-parametric template approach for reference patterns, a wide variety of tests of the recognizer have been performed with isolated word inputs in both speaker dependent (SD) and speaker independent (SI) modes. Vocabulary sizes have ranged from as small as 10 words (i.e. the digits zero-nine) to as many as 1109 words.

Table I gives a summary of recognizer performance under the conditions discussed above. It can be seen that the resulting error rates are not strictly a function of vocabulary size, but also are dependent on vocabulary complexity. Thus a simple vocabulary of 200 polysyllabic Japanese city names had a 2.7% error rate (in an SD mode), whereas a complex vocabulary of 39 alphadigit terms (in both SD and SI modes) had error rates of on the order of 4.5 to 7.0%.

Table I also shows that in cases where the same vocabulary was used in both SD and SI modes (e.g. the alphadigits and the airline words), the recognizer gave reasonably comparable performances. This result indicates that the SI mode clustering analysis, which yielded the set of SI templates or models, was capable of providing the same degree of representation of each vocabulary word as either casual or robust training for the SD mode. Sometimes the computation of the SI mode recognizer was higher than that required for the SD mode whenever a larger number of templates were used in the pattern similarity comparison.

Vocabulary	Mode	Error Rate (%)
10 Digits	SI	0.1
37 Dialer Words	SD	0
39 Alphadigits	SD	4.5
	SI	7.0
54 Computer Terms	SI	3.5
129 Airline Words	SD	1.0
	SI	2.9
200 Japanese Cities	SD	2.7
1109 Basic English	SD	4.3

Table I

**Performance of Template-Based
Isolated Word Systems**

The results in Table I are based on using word patterns (either templates or statistical models) created from isolated word training tokens. Studies have shown that when adequate training data is available, the performance of isolated word recognizers based on statistical models is comparable to or better than that of recognizers based on templates. The main issue here is the amount of training data available relative to the number of parameters to be estimated in the statistical model. For small amounts of training data, very unreliable parametric estimates result, and the template approach is sometimes superior to the statistical model approach. For moderate amounts of training data, the performance of both types of models is comparable. However, for large amounts of training data, the performance of statistical models is generally superior to that of template approaches because of their

120

ability to accurately characterize the tails of the distribution (i.e. the outliers in terms of the templates).

CONNECTED WORD RECOGNITION MODEL

The basic approach to connected word recognition from discrete reference patterns is shown in Fig. 5. Assume we are given a test pattern **T**, which represents an unknown spoken word string, and we are given a set of V reference patterns, $\{R_1, R_2, ..., R_V\}$ each representing some word of the vocabulary. The connected word recognition problem consists of finding the "super" reference pattern, $\mathbf{R}^s$, of the form

$$\mathbf{R}^s = R_{q(1)} \oplus R_{q(2)} \cdots R_{q(L)}$$

which is the concatenation of L reference patterns, $R_{q(1)}, R_{q(2)}, ..., R_{q(L)}$, which best matches the test string, **T**, in the sense that the overall distance between **T** and $\mathbf{R}^s$ is minimum over all possible choices of L, $q(1), q(2), ..., q(L)$, where the distance is an appropriately chosen distance measure.

CONNECTED WORD RECOGNITION FROM WORD TEMPLATES

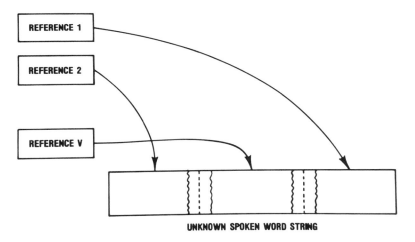

UNKNOWN SPOKEN WORD STRING

Figure 5. Illustration of Connected Word Recognition from Word Templates.

There are several problems associated with solving the above connected word recognition problem. First we don't know L, the number of words in the string. Hence our proposed solution must provide the best matches for all reasonable values of L, e.g. $L = 1, 2, ..., L_{MAX}$. Second we don't know nor can be reliably find word boundaries, even when we have postulated L, the number of words in the string. The implication is that the word recognition algorithm must work without direct knowledge of word boundaries; in fact the estimated word boundaries will be shown to be a byproduct of the matching procedure. The third problem with a template matching procedure is that the word matches are generally much poorer at the boundaries than at frames within the word. In general this is a weakness of word matching schemes which can be somewhat alleviated by the matching procedures which can apply lessor weight to the match at template boundaries than at frames within the word. A fourth problem is that word durations in the string are often grossly different (shorter) than the durations of the corresponding reference patterns. To alleviate this problem one can use some time prenormalization procedure to warp the word durations accordingly, or rely on reference patterns extracted from embedded word strings. Finally the last problem associated with matching word strings is that the combinatories of matching strings exhaustively (i.e. by trying all combinations of reference patterns in a sequential manner) is prohibitive.

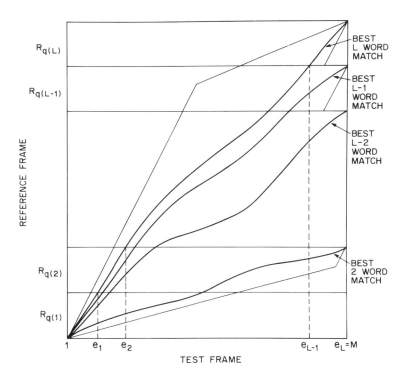

Figure 6.   Sequence of DTW Warps to Provide Best Word Sequences of Several Different Lengths.

A number of different ways of solving the connected word recognition problem have been proposed which avoid the plague of combinatorics mentioned above. Among these algorithms are the 2-level DP approach of Sakoe [11], the level building approach of Myers and Rabiner [12], the parallel single stage approach of Bridle et al. [13], and the nonuniform sampling approach of Gauvain and Mariani [14]. Although each of these approaches differs greatly in implementation, all of them are similar in that the basic procedure for finding $\mathbf{R}^s$ is to solve a time-alignment problem between $\mathbf{T}$ and $\mathbf{R}^s$ using dynamic time warping (DTW) methods.

The level building DTW based approach to connected word recognition is illustrated in Fig. 6. Shown in this figure are the warping paths for all possible length matches to the test pattern, along with the implicit word boundary markers $(e_1, e_2, ..., e_{L-1}, e_L)$ for the dynamic path of the L-word match. The level building algorithm has the property that it builds up all possible L-word matches one level (word in the string) at a time. For each string match found, a segmentation of the test string into appropriate matching regions for each reference word in $\mathbf{R}^s$ is obtained. In addition, for every string length L, the best Q matches (i.e. the Q lowest distance L-word strings) can be found. The details of the level building algorithm are available elsewhere [12], and will not be discussed here.

VOCABULARY	MODE	WORD ACCURACY	TASK	STRING (TASK) ACCURACY
Digits (10 words)	Speaker Dependent or Speaker Independent	99.8% SI	1-7 Digit Strings	99.2% SI*
		99.9% SD	1-7 Digit Strings	99.6% SD*
Letters of the Alphabet (26 words)	Speaker Dependent or Speaker Independent	≈ 90%	Directory Listing Retrieval (17,000 Name Directory)	96% SD 90% SI
Airline Terms (129 words)	Speaker Dependent or Speaker Independent	99.9% SD 97% SI	Airline Information and Reservations	99% SD 90% SI

* Known string length.

Table II

Performance of Connected Word Recognizers on Specific Recognition Tasks

Typical performance results for connected word recognizers, based on a level building implementation, are shown in Table II. For a digits vocabulary, string accuracies greater than 99% have been obtained. For name retrieval, by spelling, from a 17,000 name directory, string accuracies of from 90% to 96% have been obtained. Finally, using a moderate size vocabulary of 127 words, the accuracy of sentences for obtaining information about airlines schedules is between 90% and 99%. Here the average sentence length was close to 10 words. Many of the errors occurred in sentences with long strings of digits.

CONTINUOUS, LARGE VOCABULARY, SPEECH RECOGNITION

The area of continuous, large vocabulary, speech recognition refers to systems with at least 1000 words in the vocabulary, a syntax approaching that of natural English (i.e. an average branching factor on the order of 100), and possibly a semantic model based on a given, well defined, task. For such a problem, there are three distinct sub-problems that must be solved, namely choice of a basic recognition unit (and a modelling technique to go with it), a method of mapping recognized units into words (or, more precisely, a method of scoring words from the recognition scores of individual word units), and a way of representing the formal syntax of the recognition task (or, more precisely, a way of integrating the syntax directly into the recognition algorithm).

For each of the three parts of the continuous speech recognition problem, there are several alternative approaches. For the basic recognition unit, one could consider whole words, half syllables such as dyads, demisyllables, or diphones, or sound units as small as phonemes or phones. Whole word units, which are attractive because of our knowledge of how to handle them in connected environments, are totally impractical to train since each word could appear in a broad variety of contexts. Therefore the amount of training required to capture all the types of word environments is unrealistic. For the sub-word units, the required training is extensive, but could be carried out using a variety of well known, existing training procedures. A full system would require between 1000 and 2000 half syllable speech units. For the phoneme-like units, only about 30-100 units would have to be trained.

The problem of representing vocabulary words, in terms of the chosen speech unit, has several possible solutions. One could create a network of linked word unit models for each vocabulary word. The network could be either a deterministic (fixed) or a stochastic structure. An alternative is to do lexical access from a dictionary in which all word pronunciation variants (and possibly part of speech information) are stored, along with a mapping from pronunciation units to speech representation units.

Finally the problem of representing the task syntax, and integrating it into the recognizer, has several solutions. The task syntax, or grammar, can be represented as a deterministic state diagram, as a stochastic model (e.g. a model of word trigram statistics), or as a formal grammar. There are advantages and disadvantages to each of these approaches.

To illustrate the state of the art in large vocabulary speech recognition, consider the results shown in Table III. The table shows results for two tasks, namely an office dictation system [16], and a naval resource management system [17-22]. The office dictation system uses phoneme-like units in a statistical model to represent words, where each phoneme-like unit is a statistical model based on vector-quantized spectral outputs of a speech spectrum analysis. A third statistical model is used to represent syntax; thus the recognition task is essentially a Bayesian optimization over a triply embedded sequence of statistical models. The computational requirements are very large, but a system has been implemented using isolated word inputs for the task of automatic transcription of office dictation. For a vocabulary of 5000 words, in a speaker trained mode, with 20 minutes of training for each talker, the average *word* error rates for 5 talkers are 2% for prerecorded speech, 3.1% for read speech, and 5.7% for spontaneously spoken speech [16]. The naval resource management system also uses a set of about 2000 phoneme-like units (PLU's) to represent words where each PLU is a statistical model of a phoneme in specified contexts. The task syntax (that of a ships database) is specified in the form of a finite state network with a word pair grammar with average word branching factor (perplexity) of 60. When tested on a vocabulary of 991 words, in a speaker independent mode, using continuous, fluently spoken, sentences, a word error rate of about 4.5% was obtained [21].

Task	Syntax	Mode	Vocabulary	Word Error Rate
Office Dictation (IBM)	Word Trigram (Perplexity = 100)	SD, Isolated Word Input	5000 Words	2% – prerecorded speech 3.1% – read speech 5.7% – spontaneous speech
Naval Resource Management (DARPA)	Finite State Grammar (Perplexity = 60)	SI, Fluent Speech Input	991 Words	4.5%

Table III

Performance of Large Vocabulary Speech Recognizers on Specific Recognition Tasks

SUMMARY

In this paper we have reviewed and discussed the general pattern recognition framework for machine recognition of speech. We have discussed some of the signal processing and statistical pattern recognition aspects of the model and shown how they contribute to the recognition.

The challenges in speech recognition are many. As illustrated above, the performance of current systems is barely acceptable for large vocabulary systems, even with isolated word inputs, speaker training, and favorable talking environment. Almost every aspect of continuous speech recognition, from training to systems implementation, represents a challenge in performance, reliability, and robustness.

REFERENCES

[1] N. R. Dixon and T. B. Martin, Eds., *Automatic Speech and Speaker Recognition*, New York: IEEE Press, 1979.

[2] W. Lea, Ed., *Trends in Speech Recognition*, Englewood Cliffs, NJ: Prentice-Hall, 1980.

[3] G. R. Doddington and T. B. Schalk, "Speech Recognition: Turning Theory into Practice," *IEEE Spectrum*, Vol. 18, No. 9, pp. 26-32, Sept. 1981.

[4] L. R. Bahl, F. Jelinek, and R. L. Mercer, "A Maximum Likelihood Approach to Continuous Speech Recognition," *IEEE Trans. on Pattern Analysis and Machine Intelligence*, Vol. PAMI-5, No. 2, pp. 179-190, March 1983.

[5] A. E. Rosenberg, L. R. Rabiner, J. G. Wilpon, and D. Kahn, "Demisyllable-Based Isolated Word Recognition," *IEEE Trans. on Acoustics, Speech, and Signal Processing*, Vol. ASSP-31, No. 3, pp. 713-726, June 1983.

[6] F. Itakura, "Minimum Prediction Residual Principle Applied to Speech Recognition," *IEEE Trans. on Acoustics, Speech, and Signal Processing*, Vol. ASSP-23, pp. 67-72, Feb. 1975.

[7] F. Jelinek, "Speech Recognition by Statistical Methods," *Proc. IEEE*, Vol. 65, pp. 532-556, April 1976.

[8] S. E. Levinson, L. R. Rabiner, and M. M. Sondhi, "An Introduction to the Application of the Theory of Probabilistic Functions of a Markov Process to Automatic Speech Recognition," *Bell System Tech. Jour.*, Vol. 62, No. 4, pp. 1035-1074, April 1983.

[9] B. A. Dautrich, L. R. Rabiner, and T. B. Martin, "On the Effects of Varying Filter Bank Parameters on Isolated Word Recognition," *IEEE Trans. on Acoustics, Speech, and Signal Processing*, Vol. ASSP-31, No. 4, pp. 793-807, Aug. 1983.

[10] L. D. Markel and A. H. Gray, Jr., *Linear Prediction of Speech*, New York: Springer-Verlag, 1976.

[11] H. Sakoe, "Two Level DP Matching – A Dynamic Programming Based Pattern Matching Algorithm for Connected Word Recognition," *IEEE Trans. on Acoustics, Speech, and Signal Processing*, Vol. ASSP-27, pp. 588-595, Dec. 1979.

[12] C. S. Myers and L. R. Rabiner, "Connected Digit Recognition Using a Level Building DTW Algorithm," *IEEE Trans. on Acoustics, Speech, and Signal Processing*, Vol. ASSP-29, No. 3, pp. 351-363, June 1981.

[13] J. S. Bridle, M. D. Brown, and R. M. Chamberlain, "An Algorithm for Connected Word Recognition," *Automatic Speech Analysis and Recognition*, J. P. Haton, Ed., pp. 191-204, 1982.

[14] J. L. Gauvain and J. Mariani, "A Method for Connected Word Recognition and Word Spotting on a Microprocessor," *Proc. 1982 ICASSP*, pp. 891-894, May 1982.

[15] L. R. Rabiner, "A Tutorial on Hidden Markov Models and Selected Applications in Speech Recognition," *Proc. IEEE*, Vol. 77, No. 2, pp. 257-286, Feb. 1989.

[16] F. Jelinek, "The Development of an Experimental Discrete Dictation Recognizer," *Proc. IEEE*, Vol. 73, No. 11, pp. 1616-1624, Nov. 1985.

[17] K. F. Lee, H. W. Hon, and D. R. Reddy, "An Overview of the SPHINX Speech Recognition System," *IEEE Trans. on Acoustics, Speech, and Signal Proc.*, Vol. 38, pp. 600-610, 1990.

[18] Y. L. Chow, M. O. Dunham, O. A. Kimball, M. A. Krasner, G. F. Kubala, J. Makhoul, S. Roucos, and R. M. Schwartz, "BBYLOS: The BBN Continuous Speech Recognition System," *Proc. IEEE Int. Conf. Acoustics, Speech, and Signal Proc.*, pp. 89-92, Apr. 1987.

[19] D. B. Paul, "The Lincoln Robust Continuous Speech Recognizer," *Proc. ICASSP 89*, Glasgow, Scotland, pp. 449-452, May 1989.

[20] M. Weintraub et al., "Linguistic Constraints in Hidden Markov Model Based Speech Recognition," *Proc. ICASSP 89*, Glasgow, Scotland, pp. 699-702, May 1989.

[21] V. Zue, J. Glass, M. Phillips, and S. Seneff, "The MIT Summit Speech Recognition System: A Progress Report," *Proc. Speech and Natural Language Workshop*, pp. 179-189, Feb. 1989.

[22] C. H. Lee, L. R. Rabiner, R. Pieraccini, and J. G. Wilpon, "Acoustic Modeling for Large Vocabulary Speech Recognition," *Computer Speech and Language*, Vol. 4, pp. 127-165, 1990.

CHAPTER 6

QUALITY EVALUATION OF SPEECH PROCESSING SYSTEMS

by
Herman J.M. Steeneken
TNO-Institute for Perception
Kampweg 5
P.O. Box 23
3769 ZG Soesterberg
The Netherlands

SUMMARY

This chapter gives an overview of assessment methods for speech communication systems, speech synthesis systems and speech recognition systems. The first two systems require an evaluation in terms of intelligibility measures. Several subjective and objective measures will be discussed.

Evaluation of speech recognizers requires a different approach as the recognition rate normally depends on recognizer-specific parameters and external factors. Some results of the assessment methods for recognition systems will be discussed.

Case studies are given for each group of systems.

1 INTRODUCTION

Assessment methods for speech processing systems can be divided into three groups:

- subjective and objective intelligibility measures for speech transmission and coding systems (human-to-human)
- subjective and objective quality measures for speech output systems (machine-to-human)
- (predictive) assessment methods for speech input systems (human-to-machine).

Several methods are used for the subjective evaluation of speech transmission systems. The difference between the methods concerns mainly the type of speech material used for the test and the response method. Frequently used methods are based upon segmental evaluation, suprasegmental evaluation or overall quality measures. The linguistic units involved are phonemes, words and sentences respectively.

Objective methods, in which the transmission quality is quantified by physical parameters, are also used. The relation between these methods and their specific aspects will be discussed.

For speech output systems some additional aspects, such as intonation, may be involved. The speech signal may be synthesized from individual speech tokens such as phonemes, diphones, or larger portions, and this may result in distortions not usual for transmission channels. Some tests, including quality ratings, will be discussed.

Speech input systems are normally evaluated in relation to a specific application. This is done with a representative speech data-base or in a field experiment. However, more generally applicable evaluation methods, such as predictive methods, are also becoming available.

For all three groups the (military) application requires a careful evaluation of the environmental conditions, which may involve high noise levels, g, stress, mask microphones etc. The general approach of including these conditions in the test method will be discussed.

In some countries, such as the UK, France and the Netherlands, joint national research programs have been set up to coordinate research efforts. A European research project (sponsored by ESPRIT) was started in 1988. With this ESPRIT SAM project (multilingual speech input/output assessment, methodology and standardization) seven countries work together on the development and evaluation of speech input/output assessment methods. In the USA an advanced research program is proceeding on the development and application of speech input/output systems in military conditions. In NATO a research study group (AC/243(Panel 3)/RSG.10) is involved with the application of speech input/output systems in the multilingual military environment.

2.1 SUBJECTIVE AND OBJECTIVE INTELLIGIBILITY MEASURES FOR SPEECH TRANSMISSION AND CODING SYSTEMS (HUMAN-TO-HUMAN)

A number of subjective tests was developed during the forties, and are extensively used for the evaluation of speech communication channels. There are also two objective test methods, based on the generation and analysis of a special speech-like test signal.

We can classify the intelligibility tests with respect to their use: items tested, diagnostic information, minimum number of subjects required for reliable results, training and measuring time. Another aspect is the application: are we comparing and *rank-ordering* systems, are we evaluating a system for a *specific application* or are we supporting the *development* of a system?

When we restrict ourselves to the subjective tests, a general qualification can be made to the items tested and the response procedure. The lowest level (segmental evaluation, i.e. phonemes) is covered by the rhyme tests and the open response word tests.

A rhyme test is a multiple choice test where a listener has to select the auditorily presented word from a small group of visually presented possible responses. In general only the initial consonants of the response words are changed such as for the plosives Bam, Dam, Pam, Tam, Kam. Frequently used rhyme tests are the Diagnostic Rhyme Test (DRT) and the Modified Rhyme Test (MRT).

The DRT is based on two forced-choice alternatives [35,18,24], while the MRT is based on six alternatives [8]. As the response set is limited, a listener's response may not coincide with what is actually heard by the listener. Recent studies have shown that results obtained with a DRT may over-estimate speech intelligibility and distort the perceptual space and therefore the diagnostic value of the results [6,27].

A more general approach is obtained with an open response, as with word tests.

Word tests are based on short nonsense or meaningful words of the CVC-type (consonant-vowel-consonant). Sometimes only CV-words are used. The test words are presented in isolation or in a carrier phrase. The listener can respond with any CVC combination he has heard. Hence all confusions between the phonemes are possible. The test results include the phoneme score, the word score and the confusions between the initial consonants, vowels and final consonants. The confusion matrices present useful information to improve the performance of a system [25].

Quality rating is a more general method used to evaluate the user's opinion of a transmission channel or speech output system. Some investigators [Goodman and Nash, 5] claim that a quality rating includes the total auditory impression of speech on a listener and can be used to discriminate between good and excellent quality. For quality ratings normal test sentences or a free conversation are used to obtain the listener's impression. The listener is asked to rate his impression on a subjective scale such as the five-point scale: bad, poor, fair, good and excellent. Different types of scales are used such as: intelligibility, quality, acceptability, naturalness etc. Quality rating or the so-called Mean Opinion Score (MOS) give a wide variation among listener scores [27]. The MOS does not give an absolute measure, as the scales used by the listeners are not calibrated. Therefore, the use of reference conditions as calibration points is recommended.

The speech reception threshold (SRT) measures the word or sentence intelligibility against a level of masking noise. The listener has to recall a word or sentence masked by noise. After a correct response the noise level is increased, while after a false response the noise level is decreased. This procedure leads to an estimation of the noise level where a 50% correct recall of the presented words or sentences is obtained [19]. The quality of the speech is related to the amount of noise necessary for the masking. The procedure has the advantage that it can be performed with naive listeners.

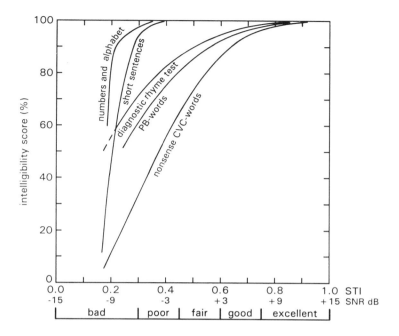

Fig. 1 Rating of some intelligibility measures and their relation with signal-to-noise ratio (SNR) for noise with a spectrum shaped according to the long-term speech spectrum.

A recent new development is the use of anomalous sentences. These are syntactically correct but semantically anomalous sentences around seven words long. The words are common mono-syllabic words from which an unlimited number of sentences can be generated randomly, according to some predefined grammatical structures.

Fig. 1 gives the score as a function of the signal-to-noise ratio of speech combined with noise for five intelligibility measures [23]. This provides an impression of the effective range of each test. The relation shown between intelligibility and the signal-to-noise ratio is valid only for noise with a power spectrum equal to the long-term speech spectrum, as is the case, for instance, with voice babble. A signal-to-noise ratio of 0 dB means that the speech and the noise have equal energy.

As can be seen from the figure the nonsense CVC-words discriminate over a wide range while meaningful test words have a slightly smaller range [1]. The digits and the alphabet give a saturation at a SNR of -5 dB. This is due to: (a) the limited number of test words and (b) recognition of these words is controlled mainly by the vowels rather than the consonants. Vowels have an average level approximately 10 dB above the average level of consonants, and are therefore more resistant to noise. On the other hand, non-linear distortion such as clipping will have a greater impact on the vowels than on the consonants. Therefore the use of the digits and the so-called phonetic alphabet, where the recognition is based mainly on vowels, may lead to misleading results. This is shown in Fig. 2, where the initial consonant score is plotted against the vowel score obtained from CVC-word tests for 68 different communication channels. The graph shows that a high vowel score and a low consonant score can be obtained on one type of channel (band-pass limiting) while the opposite can be obtained on another type of channel (e.g. peak clipping). This indicates that the exclusive use of either consonants or vowels in a subjective test leads to an incorrect evaluation of the transmission quality. A combination of consonants and vowels, as with CV or CVC words, is preferred.

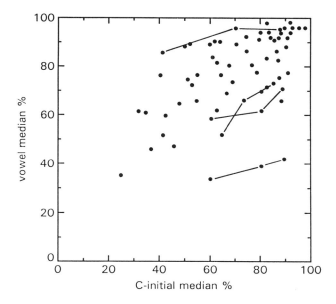

Fig. 2 Initial consonant score versus vowel score obtained from CVC-words for 68 communication channels with various combinations of bandwidth, noise, and signal-to-noise ratios. The connected datapoints have the same type of bandpass limiting, but an increasing signal-to-noise ratio.

Our research has shown that the CVC-word test based on nonsense words with the test words embedded in a carrier phrase constitutes a well-balanced test. A carrier phrase (which is neglected in many studies) will cause echoes and reverberation in conditions with distortion in the time domain. Also, AGC settling will be established by the carrier phrase. Another advantage of using a carrier phrase is that it stabilizes the vocal effort of the talker during the pronunciation and minimizes the vocal stress on the test words.

The use of nonsense words increases the open response design of such a test and extends the range of the test to higher qualities (see Fig. 1).

The reproducibility of a test strongly depends on the number of talkers and listeners used for the experiments. In general, for CVC tests, 4-8 talkers

134

and 4-8 listeners are used. It has been found that the amount of variance among individual results is equal for talkers and listeners, so in a balanced experiment these numbers should be equal.

The test-retest reproducibility can be quantified by an index (Cronbach α). This α-index gives the ratio of the within-test to between-test variances when a test is repeated. A perfect reproduction is obtained with an index of $\alpha = 1$, and poor reproduction corresponds to a small value of α.

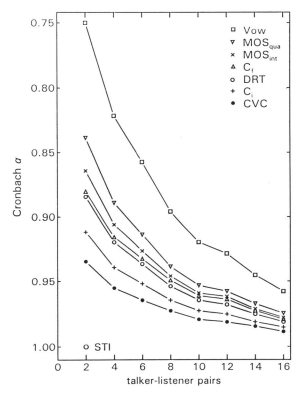

Fig. 3 Test-retest α-index as a function of the number of talker-listener pairs for some intelligibility measures (vowels, Mean Opinion Scores based on quality and intelligibility, initial and final consonants, diagnostic rhyme test and CVC-word scores).

Fig. 3 gives the α-index as a function of the number of talker-listener pairs for some of the intelligibility tests discussed above [26]. The graph shows that the index increases as a function of the number of talker-listener pairs. In general the more complex a test is (from simple vowels through to nonsense words), the more reproducible the test results are (close to $\alpha = 1$). A CVC test with four talker-listener pairs give the same α-index as a DRT with nine talker-listener pairs. Some diagnostic information can be obtained from the individual consonant and vowel scores and also from the confusions among the consonants and vowels. Multidimensional scaling techniques may help to visualize the relation between the stimuli. An example of this technique is given in [25,27] and in section 2.2, where this technique is applied to some narrowband speech coding techniques.

The effort required and the poor information on the type of degradation of the channel provided by the subjective methods have led to the development of objective measuring techniques. French and Steinberg [3] published a method for predicting the speech intelligibility of a transmission channel from its physical parameters. By using this method a relevant index (the Articulation Index, AI) was obtained. The method was reconsidered by Kryter [12], who greatly increased its accessibility by the introduction of a calculation scheme, work sheets, and tables. The AI is based on: (a) the calculation of the *effective* signal-to-noise ratio within a number of frequency bands, (b) the contribution of auditory masking, (c) a linear transformation of the effective signal-to-noise ratio to an octave-band-specific contribution from one to zero, and (d) the calculation of a weighted mean of the contributions of all octave bands considered. This method works with a calculation scheme and accounts for distortion in the frequency domain as band-pass limiting and noise. It is not applicable to distortion in the time domain and to non-linear distortion.

A method developed by Steeneken and Houtgast [23] is based on the assumption that transmission quality is closely related to the capacity of a channel to reproduce the original sound spectrum.

This can be expressed by the signal-to-noise ratio in a number of relevant frequency bands, similar to the AI approach but the method used for the measurements determines this signal-to-noise ratio dynamically. The test signal is designed in such a way that distortions in the frequency domain (non-linearities), and distortions in the time domain (echoes, reverberation, AGC) are accounted for correctly. The result is expressed with one single index, the

Speech Transmission Index (STI).

The measuring procedure is given in Fig. 4. The test signal consists of noise with a spectrum equivalent to the long-term speech spectrum (top left of Fig. 4). The intensity of each octave band is successively modulated with a sinusoid. Additive noise or noise due to distortion is added to the test signal by the effects of the transmission channel. The intensity envelope of the resulting signal contains the original modulation signal, but the depth of modulation is reduced by the distortions.

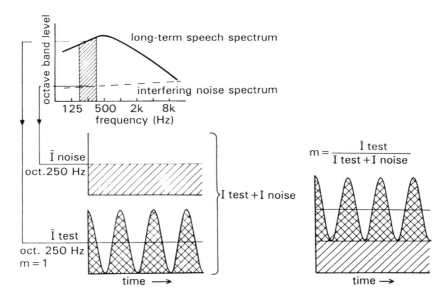

Fig. 4 Measuring scheme for obtaining the modulation transfer index.

The resulting modulation index is a measure of the signal-to-noise ratio. The modulating signals for the seven octave bands are uncorrelated (different modulation frequencies). This has the advantage that non-linear distortion components, originating in other frequency bands, are treated as noise. The signal-to-noise ratio is measured as a function of the modulation frequency, which results in the modulation transfer function. The range of the modulation frequencies corresponds to the envelope spectrum of speech (0.63-12.5 Hz).

A careful design of the characteristics of the test signal and of the type of signal analysis makes this approach widely applicable. It has been verified experimentally that a given STI implies a given effect on speech intelligibility, irrespective of the nature of the actual disturbances (noise interference, band-pass limiting, peak clipping, reverberation, etc.). The evaluation of the STI-method was performed for Dutch CVC nonsense words [23]. Anderson and Kalb [1] found similar results for English.

In Fig. 1 the rating and the relation between the STI, the signal-to-noise ratio for speech-like noise, and some subjective measures are given. The rating was obtained from an international experiment with eleven different participating laboratories [9,11].

This international evaluation resulted in a recommendation by IEC [11]. In this recommendation the RASTI-method (an application of the STI-method in room acoustics) is discussed.

On the basis of this method, a measuring device has been developed for determining the quality of speech communication systems. It comprises two parts:

1) a signal source, which replaces the talker, producing an artificial speech-like test signal, and

2) an analysis part, which replaces the listener, in which the signal at the receiving end of the system under test is evaluated.

The STI measuring equipment, which originally used special hardware, will shortly be programmed in a digital signal processor system. The use of an artificial mouth and an artificial ear (a small microphone near the entrance of the ear canal of a subject) extends the method to evaluate electro-acoustic transducers such as microphones and telephones.

2.2 APPLICATION EXAMPLES

In this section we will give four examples of the evaluation of a transmission system. Two examples based on a subjective evaluation of narrow-band secure voice systems and two examples with the objective STI: applied on a CVSD-based radio link, and to evaluate the performance of a boom microphone for use in a helicopter.

- Narrow-band secure voice terminal

A narrow-band voice terminal is usually based on a vocoder. This means that the speech signal is analyzed at the transmission side in such a way that a significant data reduction is achieved. A useful method is to determine the power spectrum and the fundamental frequency at the transmission end, typically every 20 ms, and use this information for resynthesis of the signal at the receiving end. Hence no waveform is transmitted, rather, compact information to describe the speech signal. During transmission, errors can occur concerning the spectral reproduction, the voiced/unvoiced decision, and the fundamental frequency estimation. The latter two distortions exclude the use of the existing objective measures for assessment, and hence a subjective method has to be used. A frequently used method for the evaluation is the diagnostic rhyme test DRT. A better method is the CVC test with open-response scoring. In Table I the results for two LPC systems (A, B) and a reference channel are given according to Greenspan et al. [6]. It is obvious that the rank order between the systems based on the DRT results differs from the initial consonant results and the subjective opinion scores. Greenspan showed that this could be explained by the restrictions of the DRT concept. He also showed that the differences between the scores in relation to the standard error are very small and do not allow for the conclusion that the coders are significantly different.

Table I DRT score, initial consonant scores (C_i) and subjective judgement of one reference channel and two LPC-based coders [5] (mean = m, standard error = s.e).

Coder	DRT score		C_i score		Subj. judgement	
	m	s.e	m	s.e	m	s.e
Filtered, but Uncoded	95.7%	0.77	93.8%	0.32	68.0%	2.2
Coder A	94.3%	0.62	78.6%	0.58	54.2%	1.5
Coder B	93.1%	0.64	81.6%	0.54	53.5%	1.6

- *Comparison of two LPC systems*

For two LPC systems (system A: LPC-12 4800 bps, system B: LPC-10 800 bps) the CVC-word score was measured for various transmission conditions including male/female speech, noise masking, lowered input level etc. We will give some results for the male talkers and for the condition without additive noise. The word score, the initial and final consonant score, and the vowel score are given in Table II. As an example the confusion matrix for the initial consonants of system B is also given (note the high number of confusions between f → v, R → l, b → w, and z → j, for the Dutch language).

A multidimensional scaling was performed using two confusion matrices for the two systems. The results are given in Fig. 5a,b. Fig. 5a is a presentation of the perceptual distances between the stimuli in the stimulus space. This space is based on the mean of the two input matrices. The disposition of the stimuli indicates that dimension 1 is concerned with fricatives versus plosives and dimension 2 with the voiced/unvoiced distinction. The relation in this representation of the individual coders is given in the subject space (Fig. 5b). A low loading for a particular dimension means that the stimulus space for this coder is compressed.

Table II Word and phoneme scores for two LPC-based coders and the confusion matrix for initial consonants of coder B. The results are based on 16 talker-listener pairs.

Percentage correct	C	V	C	Word
Coder A:				
Mean	69.9	75.5	78.7	41.4
Standard error	3.3	1.6	1.3	3.9
Coder B:				
Mean	59.4	63.5	56.4	24.9
Standard error	1.7	1.9	1.9	3.9

Confusion matrix for initial consonants of coder B:

Response Stimulus	P	T	K	F	S	G	M	NG	N	L	R	W	J	H	B	D	V	Z	??	Tot	%
1 P	33	1	3	2	-	-	-	-	-	-	-	2	-	2	-	1	4	-	-	48	68.8
2 T	1	29	2	6	-	-	-	-	-	-	1	-	-	-	-	1	8	-	-	48	60.4
3 K	7	4	15	4	-	4	-	-	-	-	-	-	8	1	-	4	-	-	1	48	31.3
4 F	-	3	-	14	1	8	-	-	-	-	-	-	-	-	-	-	22	-	-	48	29.2
5 S	-	-	-	-	43	-	-	-	-	-	-	-	-	-	-	-	-	5	-	48	89.6
6 G	-	-	-	-	-	47	-	-	-	-	-	-	-	-	-	-	-	-	1	48	97.9
7 M	-	-	-	-	-	-	48	-	-	-	-	-	-	-	-	-	-	-	-	48	100.0
8 NG	-	-	-	-	-	-	-	-	-	-	-	-	-	-	-	-	-	-	-	0	--.-
9 N	-	-	-	-	-	-	17	-	29	-	-	-	1	1	-	-	-	-	-	48	60.4
10 L	-	-	-	-	-	-	3	-	1	31	2	3	1	4	-	-	1	2	-	48	64.6
11 R	-	-	-	-	-	3	1	-	-	21	16	1	2	3	-	1	-	-	-	48	33.3
12 W	-	-	-	-	-	-	5	-	5	1	-	28	-	6	-	2	1	-	-	48	58.3
13 J	-	-	-	1	3	-	-	-	1	5	-	-	36	1	-	-	-	1	-	48	75.0
14 H	-	-	-	-	-	10	-	-	-	2	2	3	1	30	-	-	-	-	-	48	62.5
15 B	2	-	1	-	-	-	2	-	1	1	-	14	-	3	23	-	1	-	-	48	47.9
16 D	-	-	-	-	-	1	2	-	8	-	1	6	1	1	5	23	-	-	-	48	47.9
17 V	4	-	-	6	-	7	-	-	-	-	-	7	-	2	-	-	21	1	-	48	43.8
18 Z	-	-	-	-	3	-	-	-	1	-	-	2	20	2	-	1	-	19	-	48	39.6

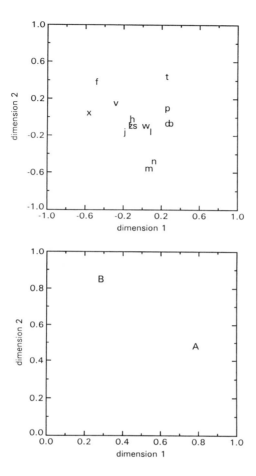

Fig. 5 Two-dimensional representation of the initial consonants passed through two LPC systems (top) and the individual weights (bottom) of the two systems in the stimulus space.

- CVSD secure voice radio link

Continuous Variable Slope Delta-modulation (CVSD) is a waveform-based coding scheme. For this reason the objective STI method can be used to determine the transmission quality. As the method gives a measuring result every 15 s, the transmission quality can be obtained as a function of the distance between an air/ground communication link. In an airplane the prerecorded STI test signal was connected to the CVSD transmission system. At a ground station a real-time analysis of the decoded signal was performed and the STI was obtained as a function of the distance between the airplane and the ground station. This measurement was performed for three types of modulation of the transmitter (base-band, diphase, and an analog reference channel) as indicated in Fig. 6. We use a criterion of an STI of 0.35 as the lower limit for a communication channel. This is comparable with 100% sentence intelligibility (Fig. 1). Based on this criterion the maximum communication distance for these conditions can be obtained from the graph and is expressed in nautical miles (23 nm, 33 nm, and 37 nm, respectively). The airplane flew at 300 ft.

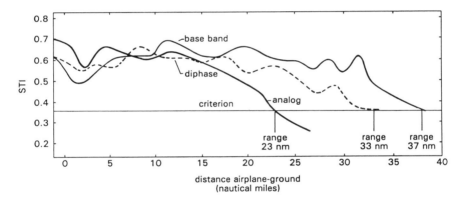

Fig. 6 Example of the STI as a function of the range for a secure CVSD radio link and an analog link between an airplane at 300 ft and a ground station.

- Microphone performance in a noisy environment
    Gradient microphones have been developed for use in high-noise environments. The specifications given by the manufacturers normally describe the effect of the noise reduction in general terms and are not related to intelligibility, microphone position or type of background noise. In Fig. 7 the transmission quality, expressed by the STI, for two types of microphones is given as a function of the environmental noise level. For these measurements an artificial head was used to obtain the test signal acoustically. The microphone was placed on this artificial head at a representative distance from the mouth. The test signal level was adjusted according to the nominal speech level. This signal level can be increased and the test-signal spectrum tilted in order to simulate the increase of the vocal effort of a talker in noise (Lombard effect). The head was placed in a diffuse sound field with an adjustable level.

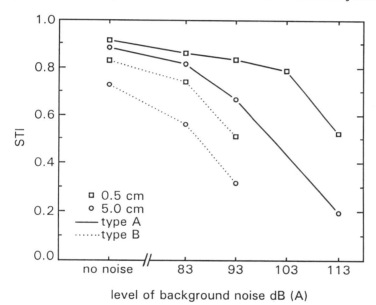

level of background noise dB (A)

Fig. 7   STI as a function of the noise level for two different microphones and two speaking distances.

From the figure we can see that the distance from the mouth is an important parameter. It is also obvious that the two noise-cancelling microphones have different performance in the noise used in this experiment.

3.1 SUBJECTIVE AND OBJECTIVE QUALITY MEASURES FOR SPEECH OUTPUT SYSTEMS (MACHINE-TO-HUMAN)

Speech synthesis has been available for more than twenty years. Starting with simple systems, which were able to reproduce short prerecorded speech tokens, the field has developed to systems converting text-to-speech.

Most speech output systems are based on waveform coding, storage and reproduction. More advanced systems are based on the coding of specific speech parameters such as spectral shape, fundamental frequency, etc. The latter method results in a more efficient coding.

Efficient coding leads to a lower speech quality but to more flexibility. A text-to-speech system can be based on elementary speech components such as phonemes or diphones. Storage or description of these elementary speech components opens the possibility of reproduction in any desired order. However, to obtain intelligible speech with an acceptable quality some other aspects have to be taken into account. For instance, the word stress, sentence accent, and the intonation contours have a major effect on the acceptability. Evaluation of speech output systems is therefore required to obtain performance figures and to obtain more diagnostic information for the improvement of the systems investigated.

The assessment method used depends on the type of system involved. For a waveform coder based on real speech tokens a segmental intelligibility test at the phoneme or word level will be satisfactory. Other acceptability items do not depend on the system itself and are defined by the speech tokens produced by a natural speaker. However, an allophone or diphone-based system also affects the intonation contours and besides a segmental intelligibility test a supra-segmental test, up to the sentence level, is required.

There are only a few experimental results on intonation. Terken and Collier [33] gave a comparison of natural speech and two synthetic intonation algorithms.

3.2 APPLICATION EXAMPLES

Recently many research results have become available for synthetic speech systems at the segmental level [20,22]. We will give here an example of the assessment by Logan et al. [11] of nine text-to-speech systems.

An example of an evaluation based on overall quality (MOS) is obtained from [21].

- Segmental intelligibility of nine text-to-speech systems

Nine text-to-speech systems were evaluated by using the Modified Rhyme Test (MRT) in two applications: with a closed-response format and with an open-response format [13]. In Table II the error rates for the nine systems and for natural speech are given. The error rates are the mean of the errors for initial and final consonants. The results show that the quality of these systems is far below the quality of natural speech. Hence natural speech can be used as a reference. No reference is used at the low quality side of the scale, though the same natural speech but at different signal-to-noise ratios could have provided such a reference. The error rate for the DEC talk for two different voice types shows that such a system also depends on the speech from which the original allophone or diphone samples were taken. The open-response format shows a wider range of scores, which is an improvement given to the similar standard error.

Table III   MRT overall error rates (% incorrect) for consonants in initial and final position [11].

System	open MRT	closed MRT
Natural speech	2.78	0.53
DECtalk 1.8 Paul	12.92	3.25
DECtalk 1.8 Betty	17.50	5.72
MITalk 79	24.56	7.00
Prose 3.0	19.42	5.72
Amiga	42.89	12.25
Infovox SA 101	37.14	12.50
Smoothtalker	56.89	27.22
Votrax Type'n'Talk	68.47	27.44
Echo	73.97	35.36

- Overall quality of three multipulse LPC-coders

Two coders (A and B) were tested in four transmission conditions: (1) nominal input level, no additive noise, (2) lowered input level, no additive noise, (3) nominal input level with additive noise, (4) lowered input level with additive noise. The Mean Opinion Score of the systems for these conditions was measured with test sentences from six male talkers. Eight listeners were asked to judge the quality on a five-point scale (1=bad, 2=poor, 3=fair, 4=good, 5=excellent). The presentation order of the sentences to the subjects was balanced so that no systematic learning effects influenced the results. The mean MOS is given in Table IV. The authors did not give the standard error, which is normally high [21]. The quality of these two coders seems to be equal. As no reference channels were used in this experiment, it is impossible to relate the scores to other experiments.

Table IV   Mean Opinion Scores for coder/speaker/condition combinations, averaged over six listeners; after [19].

Cond.	\<br\>1	\<br\>2	Speakers\<br\>3	\<br\>4	\<br\>5	\<br\>6	\<br\>Mean
Coder A:							
1	4.0	3.3	4.3	3.9	3.7	3.4	3.8
2	3.3	3.6	3.4	3.6	3.6	3.2	3.4
3	2.2	2.2	1.9	2.8	1.9	2.1	2.2
4	1.5	1.9	1.3	2.3	1.5	1.6	1.7
Coder B:							
1	3.8	4.1	3.4	4.8	3.6	3.3	3.8
2	3.3	3.7	3.7	3.2	2.9	2.5	3.2
3	2.3	2.3	2.1	3.0	1.9	1.5	2.2
4	1.5	2.0	1.3	2.3	1.0	1.4	1.7

4.1 ASSESSMENT METHODS FOR AUTOMATIC SPEECH RECOGNITION SYSTEMS (HUMAN-TO-MACHINE)

The performance of a speech recognition system depends on many parameters. The specification of the recognition performance is normally restricted to a small number of these parameters and to fixed parameter values. In general, no prediction of the performance can be made for a specific application. A more representative assessment of the system is required.

The assessment of a speech recognizer can be performed either in a field experiment with realistic conditions or in the laboratory with artificial conditions. Comparison of these two methods leads to:

field evaluation	**laboratory evaluation**
representative	artificial
uncontrolled conditions	reproducible conditions
expensive	inexpensive.

In order to gain the advantages of both methods, data-bases for representative conditions can be established (recordings of representative speech tokens) and be used many times in the laboratory. Both methods, however, have no predictive power to other applications.

This has led to different types of assessment methods, namely: data-base-oriented methods, reference-oriented methods and predictive methods.

The data-base-oriented methods normally use a custom-tailored data-base, designed for a specific application such as the use of digits, control words, single/multiple users, syntax, noise conditions, etc.

Reference-oriented methods are based on the calibration of a vocabulary or on the comparison of speech recognition systems such as proposed by:

- Moore [16] introduced the Human Equivalent Noise Ratio (HENR), a measure which expresses the performance of a recognizer with the noise level required to achieve comparable human performance.
- The Effective Vocabulary Capacity (EVC, Taylor [32]) gives the maximum size of a vocabulary that a recognizer can handle for a given tolerable error rate.
- A reference recognition algorithm is proposed by Chollet et al. [2] and later by Hieronymus [7].

The reference-based methods use a *particular* reference e.g. noise, vocabulary, or recognition algorithm. This may lead to a systematic error for a certain input variable.

Data-base-oriented and reference-oriented methods are in a very limited way diagnostic and need to be used with a representative vocabulary. Predictive methods, however, tend to be more flexible.

A method tailored to specific environmental conditions is proposed by Thomas [34]. He uses a set of small representative data-bases that have been calibrated for certain aspects. The calibration is obtained from overall measures of the speech data and may be dominated by the vowels.

Two diagnostic methods, both also used for the intelligibility evaluation, have been proposed. The Phonetic Discrimination test (PD-100) [31] is based on one hundred test words, constructed on minimally different phonetic aspects as used with the two alternative rhyme tests. Both correct responses and false responses are used. The Recognition Assessment by Manipulation Of Speech (RAMOS [29]) is not limited to single phonetic aspects between test words and uses an open response set with all possible confusions of initial consonants, vowels, and final consonants in CVC-type test words. The method is extended to connected-speech recognition by embedding the test words in a carrier phrase. The test words are analyzed and resynthesized by an algorithm. This offers the possibility of manipulating the speech before the resynthesis to model variations due to inter- and intra-speaker variations. For this purpose an analysis of speech data has been carried out. An example of the application of the method is given in the next section.

Since the results of all assessment methods are influenced by external factors, the evaluation of a recognition system must be performed under controlled and specified conditions. These factors can be divided into broad categories. Each category can be divided into specific, individual factors, such as:

- Speech   isolated words
                      connected words
                      connected discourse

- Speaker   speaker dependency
                      speaker within/outside reference patterns
                      age, sex, accent, native language
                      recording conditions
                      vocal effort
                      speaking rate
                      language dependency

- Task   size, redundancy vocabulary
                      complexity of syntax

- Environment   noise
                      reverberation, echoes
                      co-channel interference

- Input   microphone
                      system noise
                      distortion

     - Recognizer   system parameters, thresholds
                             training.

One can determine the percentage of correctly recognized words as a function of all these parameters. For many applications, however, this is not sufficient. We also need to know the number of confusions and rejections separately. For an isolated-word recognizer, for instance, the following performance measures can be determined:

     - Words inside vocab.   percentage correct
                                   percentage rejected
                                   percentage incorrect

     - Words outside vocab.   percentage rejected (which is correct)
                                   percentage incorrect (all positive responses)

     - Confusions between words inside and outside vocabulary

     - Predictive measures.

For connected-word recognizers an additional measure can be added:
     -                               percentage insertions
     -                               percentage deletions.

Hunt [10] describes a popular method of determining the percentages of insertions, deletions and substitutions using dynamic programming sequence matching to compare the sequence of words found by the recognizer to the known correct sequence. He points out, however, that the resulting percentages tend to be biased, understating insertions and deletions and overstating substitutions. A total errors measure obtained by summing the three kinds of errors will also be biased, and the amount of bias will vary from test to test, making results less reproducible. His *weighted total errors* measure, in which insertions and deletions are given a weight of a half and substitutions one, is free of this bias.

Significance of the performance of different recognizers can be tested by means of statistical tests such as the analysis of variance (ANOVA) and the

McNemar test [4]. As the test results are error rates, no straightforward ANOVA can be applied but a lin/log transformation is required. For some vocabularies a very low error rate is obtained, and in this case the application of a statistical test requires a very high number of trials to get significant results. In our opinion a more difficult vocabulary would be preferable. This is similar to the relation obtained between nonsense CVC-words and short sentences for intelligibility testing.

International standardization of assessment methods is a necessity if comparable results are to be obtained. Some years ago the NATO research study group RSG.10 established a data-base for isolated and connected digits and for native and non-native talkers. This data-base has been used for many experiments at different locations. An example of an evaluation using this data-base is given in section 4.2.

Some other bodies working on test standardization are ESPRIT-SAM and Nat. Inst. of Standards and Technology (NIST, formerly NBS). Both bodies have established speech data-bases on a CD-ROM. RSG.10 has recorded an existing noise data-base on CD-ROM [28]. Also, in order to specify signal-to-noise ratios in a reproducible manner, the standardization of speech level measures, has been adapted to a simple PC-workstation [26].

4.2 APPLICATION EXAMPLES

The recognition rate depends very much on the acceptance criterion between the fit of the best matching template and the speech token. The higher the acceptance criterion the lower the number of correct responses, the higher the number of rejections but also the lower the number of false responses. In Fig. 8 an example is given for an isolated-word recognizer trained with 68 words. The figure shows the recognition rate as a function of the acceptance threshold (solid line), the figure also gives the rate of false responses for words outside the vocabulary (this is equivalent to the score of the second choice responses). For this recognizer with this vocabulary the optimum threshold setting is around "18", where the best separation between a high correct-response score and a low false-response score is achieved.

Speech recognition for connected words and for talkers using a language other than their native language is a problem that arises in a multinational community such as NATO. RSG.10 therefore conducted an experiment with non-native talkers and with isolated and connected digits [17].

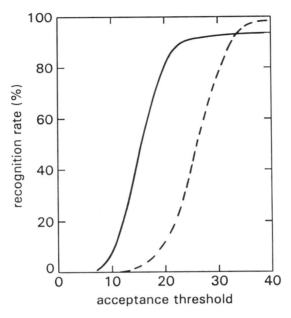

Fig. 8 Recognition rate as a function of the acceptance threshold for correct responses (solid line) and false responses (dotted line).

In Fig. 9 the error rate for five recognizers and humans is given for groups of digits of 1, 3, 4, and 5 connected digits respectively. The error rate increases with the number of digits in a group. All systems show a good performance for isolated digits. For connected digits some recognizers show poor performance.

It was found in this study that a slightly different result is obtained for male and female voices. However, some recognizers perform better with female voices, while other do better with male voices.

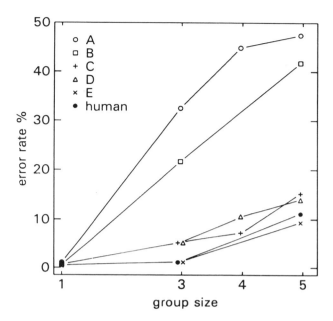

Fig. 9   Effect of group size of connected digits on the recognition error rate for five connected-speech recognizers and humans [17].

The effect of language and native/non-native talkers speaking English digits is very significant. The individual speaker variation, however, explains more variance than any other parameter. Similar results were found by Steeneken, Tomlinson and Gauvain [30].

A method in which the recognizer performance is specified as a function of the variation of specific speech parameters and environmental conditions has already been mentioned [29]. The method uses a small test vocabulary with minimal-difference word sets of CVC-type words. For the initial consonants, words such as pil, kil, til, bil, dil etc. are used; similarly, for the final consonants lip, lik, lit etc., and for vowels tat, toot, teet, tit, tot etc. Training and scoring are according to the open-response experimental design. This results in valuable diagnostic properties. By means of an analysis-resynthesis technique, the test words can be physically manipulated according to changes of human speech for well defined conditions. For this purpose a cook-book is under development which describes the relevant parameters and the amount of variation for conditions like inter- and intra-speaker variability, male/female, normal/stressed etc.

Fig. 10 shows the error rate for initial consonants with four recognizers and humans as a function of a shift in the second formant. The training of the recognizers was carried out with the original resynthesized words. The manipulation procedure has the advantage that the recognizer performance can be studied as a function of a specific parameter or combination of parameters. This is not the case with the use of a (representative) data-base. where normally one unknown condition is used.

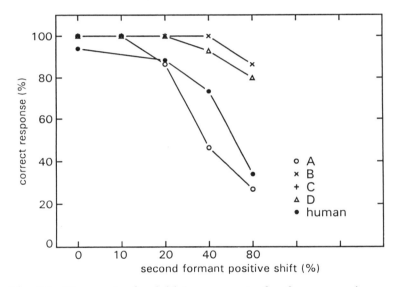

Fig. 10 Error rate for initial consonants for four recognizers and human listeners as a function of the second formant shift.

Another example is the use of noise. In Figs 11 and 12 the error rate for initial consonants is given as a function of the signal-to-noise ratio for additive noise. The superior performance of humans is obvious. Training the recognizers in noise shifts the SNR range over which adequate recognition performance can be obtained.

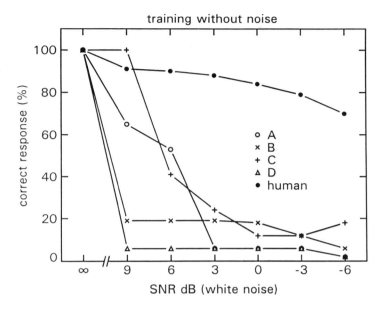

Fig. 11 Error rate for initial consonants, four recognizers and human listeners as a function of the signal-to-noise ratio. The training was performed without noise.

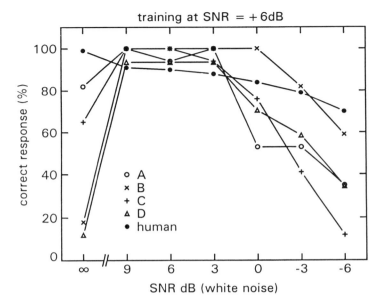

Fig. 12 Error rate for initial consonants, four recognizers and human listeners as a function of the signal-to-noise ratio. The training was performed at a signal-to-noise ratio of 6 dB.

5   FINAL REMARKS AND CONCLUSIONS

Some examples of evaluation methods of speech systems have been given. Several items for future development were identified, such as evaluation at sentence level for speech output systems and objective evaluation methods for speech input systems.

The availability of standardized evaluation methods and data-bases increases the possibility to compare results from different studies.

There is a need for diagnostic assessment methods especially to evaluate speech recognition systems. The use of multidimensional scaling techniques in combination with tests resulting in a confusion matrix based on low-level stimuli (i.e. phonemes) shows promising results.

An aspect not discussed in this review, but relevant to the application of speech input/output systems in combination with computer systems, is the dialogue structure and the human-computer interface.

ACKNOWLEDGEMENTS

I am grateful to Dr. Melvyn Hunt for his critical review of the manuscript.

6   REFERENCES

[1] Anderson, B.W. and Kalb, J.T. English verification of the STI method for estimating speech intelligibility of a communications channel. J. Acoust. Soc. Am. **81** (6), (1987), 1982-1985.

[2] Chollet, G.F., Gagnoulet, C., "On the Evaluation of recognizers and databases using a reference system", IEEE Proc. ICASSP, Atlanta (1981).

[3] French, N.R. and Steinberg, J.C., Factors governing the intelligibility of speech sounds. J. Acoust. Soc. Am. **19** (1947), 90.

[4] Gillick, L. and Cox, S.J., Some statistical issues in the comparison of speech recognition algorithms, IEEE Proc. ICASSP, Glasgow (1989).

[5] Goodman, D.J. and Nash, R.D., Subjective quality of the same speech transmission conditions in seven different countries, IEEE Trans Comm. **30** (1984) 642-654.

[6] Greenspan, S.L., Bennett, R.W. and Syrdal, A.K., A study of Two Standard Speech Intelligibility Measures. Presented 117th Meeting Acoust. Soc. Am., May 1989.

[7] Hieronymus, J.L., Majurski, W.J., "A reference speech recognition algorithm for benchmarking and speech data-base analysis", IEEE Proc. ICASSP, Tampa (1985).

[8] House, A.S., Williams, C.E., Hecker, M.H.L. and Kryter, K.D., Articulation testing Methods: Consonantal differentiation with a closed response set., J. Acoust Soc. Am. **37** (1965), 158-166.

[9] Houtgast, T. and Steeneken, H.J.M., A multilanguage evaluation of the Rasti-method for estimating speech intelligibility in auditoria. Acustica **54** (1984), 185-199.

[10] Hunt, M.J., Figures of merit for assessing connected-word recognizers. Speech Communication **9** (1990), 329-336.

[11] IEC-report. The objective rating of speech intelligibility in auditoria by the "RASTI" method, Publication IEC 268-16 (1988).

[12] Kryter, K.D., Methods for the calculation and use of the articulation index. J. Acoust. Soc. Am. **34** (1962), 1689-1697.

[13] Logan, J.S., Greene, B.G. and Pisoni, D.B., Segmental intelligibility of synthetic speech produced by rule. J. Acoust. Soc. Am. **86** (2), (1989), 566-581.

[14] Mariani, J., Covering notes concerning the survey on existing voice recognition equipments. AC/243(Panel 3) RSG-10 document (1989).

[15] Michael Nye, J., Human factors analysis of Speech Recognition systems. Speech Technology, Vol 1 (1982), No.2.

[16] Moore, R.K., Evaluating speech recognizers. IEEE Trans. ASSP, Vol ASSP-25, No. 2 (1977), 178-183.

[17] Moore, R.K., Report on connected digit recognition in a multilingual environment. Report AC/243(Panel 3)D/259, January 25, 1988.

[18] Peckels, J.P. and Rossi, M., Le test diagnostic par paires minimales. Revue d'Acoustique No 27 (1973), 245-262.

[19] Plomp, R. and Mimpen, A.M., Improving the reliability of testing the

speech reception threshold for sentences, Audiology **8** (1979), 43-52.

[20] Pols, L.C.W., Improving synthetic speech quality by systematic evaluation. Proceedings ESCA workshop, Noordwijkerhout (1989), The Netherlands.

[21] Son, N. van, and Pols, L.C.W., "Final evaluation of three multipulse LPC coders: CVC intelligibility, Quality Assessment and speaker identification." Report IZF 1989-17 (1989), TNO Institute for Perception, Soesterberg, The Netherlands.

[22] Spiegel, M., Altom, M.J., Macchi, K. and Wallace, K., A monosyllabic test corpus to evaluate the intelligibility of synthesized and natural speech. Proceedings ESCA Workshop, Noordwijkerhout (1989), The Netherlands.

[23] Steeneken, H.J.M. and Houtgast, T., A physical method for measuring speech-transmission quality. J. Acoust. Soc. Am. **67** (1), (1980), 318-326.

[24] Steeneken, H.J.M., Ontwikkeling en toetsing van een Nederlandstalige diagnostische rijmtest voor het testen van spraakkommunikatiekanalen. Report IZF 1982-13 (1982), TNO Institute for Perception, Soesterberg, The Netherlands.

[25] Steeneken, H.J.M., Diagnostic information of subjective intelligibility tests. Internat. IEEE Proc., ICASSP, Dallas (1986).

[26] Steeneken, H.J.M. and Houtgast, T., Comparison of some methods for measuring speech levels. Report IZF 1986-20 (1986), TNO Institute for Perception, Soesterberg, The Netherlands.

[27] Steeneken, H.J.M., Comparison among three subjective and one objective intelligibility test. Report IZF 1987-8 (1987), TNO Institute for Perception, Soesterberg, The Netherlands.

[28] Steeneken, H.J.M. and Geurtsen, F.W.M., Description of the RSG-10 Noise Data-base. Report IZF 1988-3 (1988), TNO Institute for Perception, Soesterberg, The Netherlands.

160

[29] Steeneken, H.J.M. and Van Velden, J.G., Objective and diagnostic assessment of (isolated) word recognizers. IEEE Proc. ICASSP, Glasgow (1989), 540-543.

[30] Steeneken, H.J.M., Tomlinson, M., and Gauvain, J.L., Assessment of two commercial recognizers with the SAM workstation and Eurom 0. Proceedings ESCA workshop, Noordwijkerhout (1989), The Netherlands.

[31] Simpson, C.A., and Ruth, J.C., "The Phonetic Discrimination Test for Speech Recognizers", Part I, Speech Technology, March/April 1987, Part II, Speech Technology Oct/Nov 1987.

[32] Taylor, M.M., "Issues in the evaluation of speech recognition systems", J. Am. Voice I/O Soc., Vol 3 (1986), 34-68.

[33] Terken, J.M.B. and Collier, R., Automatic synthesis of natural-sounding intonation for text-to-speech conversion in Dutch. Proceedings Eurospeech 89, Paris (1989), September 26-28.

[34] Thomas, T.J., "The prediction of speech recognizer performance by the use of laboratory experiments: some preliminary observations", Proc. European Conf. on Speech Technology, Edinburgh (1987), Vol 2, 245-248.

[35] Voiers W.D., Diagnostic Evaluation of Speech Intelligibility. Chapter 32 in M.E. Hawley (ed.) Speech Intelligibility and speaker recognition, Vol. 2. Benchmark papers in Acoustics, Dowden, Hutchinson, and Ross, (1977), Stroudburg, Pa.

CHAPTER 7

SPEECH PROCESSING STANDARDS

A. Nejat Ince

Istanbul Technical University
Ayazaga Campus
Istanbul
Turkey

ABSTRACT

This chapter deals with speech processing standards for 64, 32, 16 kb/s and lower rate coders and more generally, speech-band signals which are or will be promulgated by CCITT and NATO. The International Telegraph and Telephone Consultative Committee (CCITT) deals, among other things, with speech processing within the context of ISDN. Within NATO there are also bodies promulgating standards which make interoperability possible without complex and expensive interfaces.

The chapter highlights also some of the applications for low-bit rate coding and the related work undertaken by CCITT Study Groups which are responsible for developing standards in terms of encoding algorithms, codec design objectives as well as standards on the assessment of speech transmission quality.

1. STANDARDS ORGANISATIONS

The dictionary meaning of "Standards" as applied to Telecommunications is "Something established by authority, custom or general consent as a model or example". Standards are known by different names depending on the source, for example, standards, specifications, Regulations, Recommendations. By any name, their purpose is to achieve the necessary or desired, degree of uniformity in design or operation to permit systems to function beneficially for both providers and users. The intended scope of standards can vary. They may be internal within a company or they may apply to an entire country, a world region, or the world as a whole.

This chapter deals only with international (global and regional CEPT and NATO) standards as far as speech processing systems are concerned.

International Standards organisations are of two types, treaty based and voluntary.

The treaty based world organisation is the International Telecommunication Convention, a multilateral treaty. The International Consultative Committee for Telegraph and Telephone (CCITT) and the International Consultative Committee for Radio (CCIR) are the two technical organs of the ITU involved in standards making.

The standardization activities of the Conference of European Posts and Telecommunications Administrations (CEPT) supplement the action of the CCITT. Generally a preliminary agreement among European countries enables common proposals to be introduced which make the work of CCITT study groups easier and quicker. In other cases, where international Recommendations offer several choices, the CEPT encourages its members to adopt the same solution. In addition new systems jointly studied by several European countries also form the subject of Recommendations at the international level. Finally, the CEPT is to define a common system for the approval procedures applying to terminal equipment. All these activities result in the drafting of Recommendations by the CEPT. However, in terms of its activities, the CEPT never competes or opposes the action of the CCITT or any other international organisation.

Military Standards both procedures and materials required by the NATO countries to enable their forces to operate together in the most effective manner are evolved in NATO by various committees, and groups and are promulgated by the "Military Agency for Standardisation" (MAS) in the form of NATO Standardisation Agreements (STANAG's).

The so-called "Voluntary or Industry Standards" are documents prepared by nationally recognized industrial and trade associations and professional societies for use by the general public. Most of these "standards" usually feed into the work of the international standards organisations.

2. WORKING METHODS OF THE CCITT

The primary objectives of the CCITT are to standardise, to the extent necessary, techniques and operations in telecommunications to achieve end-to-end compatibility of international telecommunication connections, regardless of the countries of origin and destination. CCITT Standards are usefull also for national applications and in most countries today national and even local equipment comply with CCITT Standards. In developing Standards, the CCITT is required by its rules to invite other organisations to give specialist advice on subjects that are of mutual interest. Thus, very close cooperation is ensured and account taken of the work done by other organisations such as ISO and IEC.

The main principles of the working procedures of the CCITT are set out in the International Telecommunication Convention whereas the detailed procedures are contained in various resolutions of the CCITT Plenary Assemblies [1].

The work program of the CCITT in the various domains such as transmission and switching is established at every Plenary Assembly in the form of Questions submitted by the various Study Groups based on requests made to the Study Groups by their members. The Plenary Assembly assesses the various Study Questions, reviews the scope of the Study Groups, and allocates Questions to them. The Study Groups organise their work (that is which Questions are to be dealt with by the Plenary of the Study Group, by a working Party, a Special Rapporteurs' group or an ad hoc group) and appoint the chairman, Special Rapporteurs, etc.

Work on CCITT Study Question normaly leads to one or several draft Recommendations to be Submitted for approval to the next Plenary Assembly. All Recommendations, new or amended, are printed in the various volumes of the CCITT Book after approval.

The present CCITT Study Groups together with their areas of interest are given in Table I below.

Table I

CCITT Study Groups and Their Areas of Responsibility

Study Group	Area
I	Definition and operational aspect of telegraph and telematic services (such as facsimile, telex, and videotex)
II	Telephone operation and quality of service
III	General tariff principles
IV	Transmission maintenance of international lines, circuits, and chains of circuits; maintenance of automatic networks
V	Protection against dangers and disturbances of electro-magnetic origin
VI	Protection and specifications of cable sheaths and poles
VII	Data communications networks
VIII (XIV)	Terminal equipment for telematic services
IX (X)	Telegraph networks and terminal equipment
XI	Telephone switching and signalling
XII	Telephone transmission performance and local telephone

	networks
XV	Transmission systems
XVI	Telephone circuits
XVII	Data communications over the telephone network
XVIII	Digital networks

2.1 Procedures for Speech Processing Standardsation

In the CCITT, several Study Groups are involved in speech processing standardisation activities. Study Group XVIII's Working Party on Speech Processing is responsible for setting up the standards in terms of encoding algorithms and related codec design objectives. Standards on the assessment of speech quality fall under the responsibility of Study Group XII. Standards on network objectives, to which the codec design and performance should comply, are the subject of study for standardisation by Study Group XVIII (digital networks) and Study Group XV (mixed analog-digital networks).

The Working Party on Speech Processing of Study Group XVIII has been acting for several Study Periods (a four year time period) as the coordinating body that plans the various steps of the standardisation process addressed to both new technologies for network transport (e.g. adaptive differential pulse code modulation (ADPCM) at 32 kbits/s) and new technologies in support of new services capabilities (e.g., coding of wide-band speech i.e., 7 kHz, within 64 kbits/s).

The procedure to obtain a consensus on the standard processing algorithms consists of a technical selection among competing candidate codes. Selection is done on the basis of series of subjective (to assess speech quality) and objective laboratory tests (voiceband data quality) on prototype codecs. Tests are carried out in different world locations according to standard CCITT measurement conditions and procedures. They aim to verify codec performance under realistic network environmental conditions. Typical test conditions include:
- single encoding,
- single encoding with injected digital errors with random or bursty arrival statistics,
- synchronous and asynchronous tandem encoding for up to eight links in tandem (synchronous refers to digital-to-digital tandem encoding between 64 kbit/s PCM and another digital coding format; asynchronous tandem encodings involve analog signal representation between successive encodings),
- asynchronous tandem encoding with injected analog impairments (noise, loss, amplitude and delay distortion, phase jitter, harmonic distortion), these conditions being critical for voiceband data performance.

Voice quality is based on subjective listening tests with absolute judgment scores. This test uses a five-point scale and is based on mean opinion score (MOS) judgements under defined test conditions. These conditions include: source speech, reference conditions (white noise, speech-correlated noise, handset characteristics), digital error generation, and test administration. As can be seen in Chapter 6 of this book, MOS is not considered to be accurate and better discriminating tests are therefore being considered.

Voice band data quality is assessed in the same network environment as for voice on a variety of CCITT-specified modems and facsimile equipment. Quality is measured on the basis of bit or block error performance.

Selection activities are conducted by a group of experts consisting of representatives from administrations, operating agencies, and manufacturers, that must establish a multilaboratory test workplan, evaluate the obtained performance, and finally agree on a specific codec algorithm by taking into account other aspects such as

. codec complexity,

. codec delay,

. ease of transcoding with PCM,

. amenability to variable rate coding.

It is to be noted that standards result from the selection among competing systems, and they also incorporate some of the best features of their competitors. Another point that should be mentioned about standards is that they sometimes turn out not to be satisfactory in the field in which case they are reconsidered and modified.

3. CCITT SPEECH PROCESSING STANDARDS

Before we discuss recent and future CCITT standardisation activities we should mention that the first and the most significant milestone in speech processing standards was achieved at the end of the 1960's (amended in 1972) with the promulgation of the CCITT Recommendations G.711 concerning "Pulse Code Modulation (PCM) of voice Frequencies" [2]. This recommendation specified (together with Rec.G.702) 64 kb/s PCM coding using a sampling rate of 8000 samples per second and two encoding laws commonly referred to as the A-law and the μ-law. PCM which dominates speech processing applications in today's networks has a high degree of robustness to transmission errors and tandem encodings and offers satisfactory performance to speech and voiceband data in most mixed applications.

Advances made in previous years in digital signal processsing eventually caused in the study period 1982-84 the initiation of standardisation activites in CCITT in Speech processing [3] and the setting up, under Study Group XVIII, of a Working Party to deal with the establishment of such standards.

The standardisation activities of CCITT in speech processing may be divided into two groups:

a) those related to the so-called "low-bit-rate-voice" , (LBRV) with a bit rate less than 32 kb/s which aim at overc oming in the short-to-medium terms, before the widespread use of the emerg ing optical fibre, the economic weakness of 64 kb/s PCM in satellite and long-haul terrestrial links and copper subscriber loops, and

b) those associated with " high-fidelity voice" (HFV) with bandwidth up to 7 kHz for applications such as loudspeaker telephones, teleconferencing and commentary channels for broadcasting, within a bit rate of 64 kb/s.

The standards that have been issued and the ones on which work is still in progress are outlined below for the cases (a) and (b) above.

3.1. 32 kb/s Adaptive Differential Pulse Code Modulation (ADPCM)

The latest version of CCITT Recommendation G.721 [4,5] (first approved in 1984 and revised in 1986) specifies standards for the conversion of a 64 kb/s A-law or μ-law PCM channel to and from a 32 kb/s channel. In the ADPCM encoder, Fig. 1, the A/μ-law PCM input signal is first converted into uniform PCM and then a difference signal is obtained, $d(n)$, by subtracting an estimate of the input signal, $\hat{x}(n)$, from the input signal itself, $x(n)$. An adaptive 15-level quantiser is used to assign four binary digits to the value of the difference signal for transmission to the coder. An inverse quantiser produces a quantised difference signal for the same four digits. The signal estimate is added to the quantised difference signal to produce the reconstructed version of the input signal. Both the reconstructed signal, p_n and the quantised difference signal, $d_q(n)$, are operated upon by an adaptive predictor which produces the estimate of the input signal, $\hat{x}(n)$, thereby completing the feedback loop.

The ADPCM decoder includes a structure identical to the feedback portion of the encoder, together with a uniform PCM to A-law or μ-law conversion and a synchronous coding adjustment. The synchronous coding adjustment prevents cumulative distortion occuring on synchronous tandem codings (ADPCM-PCM-ADPCM etc. digital connections) under certain conditions. The synchronous coding adjustment is achieved by adjusting the PCM output codes in

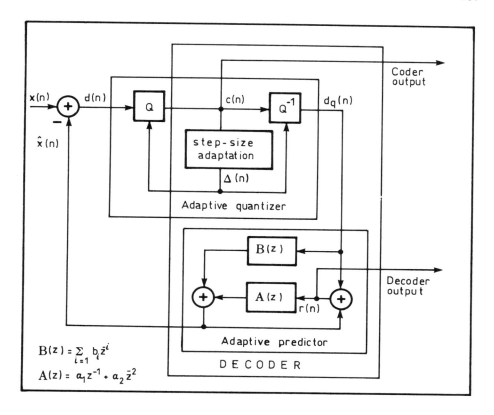

Fig.1. Block Diagram of an Adaptive Differential Pulse Code
Modulation Encoder / Decoder

a manner which attempts to eliminate quantising distortion in the next ADPCM
encoding stage.

    As can be seen in Fig. 2 the perceived quality of speech over 32 kb/s ADPCM
links is comparable to 64 kb/s PCM for up to two asynchronous codings, slightly
poorer for four codings and significantly worse with eight codings. It is clear that
the deployment of asynchronous tandem codings of ADPCM in the network must
be limited. CCITT have adopted a voice criterion which allows a maximum of four
asynchronous ADPCM codings on an end-to-end connection if there is no other
source of quantizing distortion. In addition, CCITT Recommendation G.113
allows one ADPCM coding in the national network on the national extension of an
international connection. On the other hand, ADPCM is more robust than PCM in
the presence of random bit errors.

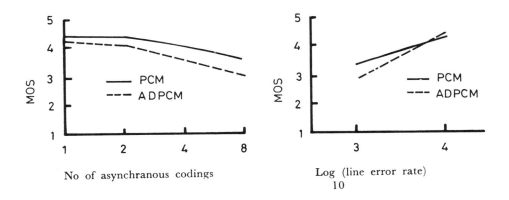

Fig.2. Subjective Performance for PCM and ADPCM Codings [5]

Block Error-rate (BLER with 1000 bits in a block) test results, carried out with random additive noise (and some other added analog impairment such as delay distortion, non-linear distortion and phase jitter), for 2400 b/s V.26 and 4800 b/s V.27 modems show that, with an acceptability criterion of a 10^{-2} BLER at an S/N of 24 dB, ADPCM provides an acceptable level of performance with both modems and with four asynchronous codings. As expected, the degredation with ADPCM relative to PCM is more pronounced with the higher speed V.27 signals. Performance of 9600 b/s V.29 is not acceptable for even one ADPCM coding.

For applications where bit error rate (BER) is recognised as an important criterion, the acceptability limit is BER $< 10^{-5}$. This is almost always a more stringent constraint than BLER. Using this criterion, some modems provide acceptable performance with only two or three asynchronous codings at the 4800 b/s rate. In general, the impact on voiceband data performance is considerable even when limiting criteria are met.

Classical transmission measurements such as S/N must be interpreted with care for adaptive signal processing algorithms such as ADPCM, since S/N typically depends on input signal statistics. In other words, such measurements, in general, cannot be used to predict performance for other signals with significantly different spectral and temporal characteristics.

3.2. 7 kHz Audio-Coding Within 64 kb/s

The CCITT Recommendation G.722 [6,7] describes the characteristics of an audio (50 to 7000 Hz) coding system which may be used for a variety of higher quality speech applications. The coding system uses sub-band adaptive differential pulse code modulation (SB-ADPCM) within a bit rate of 64 kb/s (see Fig. 3). In the technique used, the frequency band is split into two sub-bands (higher and lower) and the signals in each sub-band are encoded using ADPCM. The system has three basic modes of operation corresponding to the bit rates used for 7 kHz audio coding at 64, 56 and 48 kb/s which are the subjects of Draft Recommendation G.72y and Y.221 (Frame structure for a 64 kb/s Channel in Audio-Visual Teleservices) having other speech bit rates, or data rates up to a full 64 kb/s data path.

The 64 kb/s (7 kHz) audio encoder comprises a transmit audio part which converts the audio signal to a uniform digital signal which is coded using 14 bits with 16 kHz sampling and a SB-ADPCM encoder which reduces the bit rate to 64 kb/s.

The corresponding decoder comprises a) a SB-ADPCM decoder which performs the reverse operation to the encoder noting that the effective audio coding bit rate at the input of the decoder can be 64, 56 or 48 kb/s depending on the mode of operation; and b) a receive audio part which reconstructs the audio

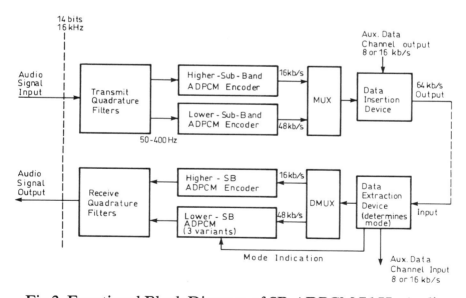

Fig.3. Functional Block Diagram of SB-ADPCM 7 kHz Audio Coding within 64 kb/s [6]

signal from the uniform digital signal which is encoded using 14 bits with 16 kHz sampling.

For applications requiring an auxiliary data channel within the 64 kb/s the following two parts are needed:

- a data insertion device at the transmit end which makes use of, when needed, 1 or 2 audio bits per octet depending on the mode of operation and substitutes data bits to provide an auxiliary data channel of 8 or 16 kb/s respectively;

- a data extraction device at the receive end which determines the mode of operation according to a mode control strategy and extracts the data bits as appropriate.

The wideband speech algorithm outlined above was selected by assuming end-to-end digital connectivity and excluding the requirement of voiceband data transmission or asynchronous tandem encodings (synchronous transcoding to and from uniform PCM to provide conference bridge arrangements is required).

To allow switching among 64, 56, and 48 kb/s speech coding rates, the lower subband (0-4000 Hz) ADPCM coder is designed to operate at 6, 5 or 4 bit/sample. Embedded coding is used to prevent quality degradation in case of a mismatched mode of operation between the encoder and decoder.

Fig. 4 shows audio quality for speech and music as a function of the G.72z algorithm bit rates. The performance of 240 kb/s linear PCM (15-bit audio input sampled at 16 kHz) is also given for comparison as well as the current realistic research goal for the coding of 7-kHz audio.

The subjective evaluation tests conducted with the standard algorithm in terms of average MOS versus the three encoding bit rates at different BER show [7] that when BER is better than 10^{-4} MOS stays around the value of 4 increasing slightly with bit rate whereas at BER $= 10^{-3}$ the MOS remains almost constant with bit rate at the value of 3. With four synchronous transcodings, MOS changes from about 3 to 4 with bit rate for BER $> 10^{-4}$ (see Fig. 5).

3.3. Draft CCITT Recommendation G.72z

This recommendation which will probably be numbered G.723, extends Rec.G.721 to include the conversion of a 64 kb/s A-law or μ-law PCM channel to and from a 24 kb/s or 40 kb/s channel [8]. The principal application of 24 kb/s channels is for overload channels carrying voice signals in Digital Circuit Multiplication Equipment (DCME) (see Fig. 6). 40 kb/s channels are used mainly

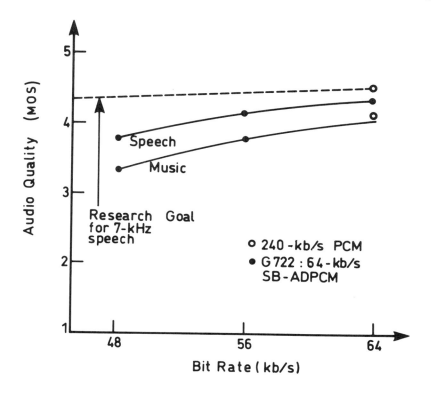

Fig.4. 7 kHz Digital Audio Quality for Speech and Music
at G.722 bit rates [16]

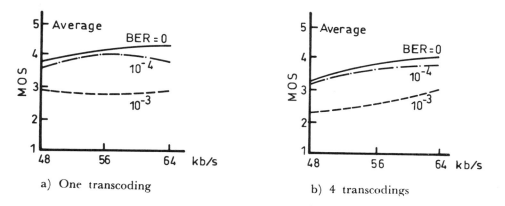

a) One transcoding

b) 4 transcodings

Fig.5. Subjective Evaluation Tests With SB-ADPCM Codec [7]

for carrying data modem signals in DCME, especially for modems operating at greater than 4800 b/s (32 kb/s channels do not perform well with 9.6 kb/s modems).

Fig.6. Point-to-Point or Circuit-Based DCMS

DCME makes use of digital speech interpolation (DSI) and low-bit-rate voice techniques to increase, with respect to 64 kb/s PCM, the number of simultaneous voice calls transmitted over a digital link [9]. DSI takes advantage of limited voice activity during a call (less than 40% of the time) and transmits only the active parts of a conversation (talkspurts). The channel capacity is allocated to talkspurts from other conversations during silent intervals. The use of variable bit rate coding of talkspurts avoids effectively the annoying "freeze-out" effect which, if allowed to occur, would result in the loss of a talkspurt as a consequence of excessive traffic load on the digital link.

G.72z recommends that when using 32 kb/s ADPCM, coding should be alternated rapidly to 24 kb/s such that at least 3.5 to 3.7 bits/sample are used on average (for further study). The effect on speech quality of this alternation is not expected to be significant. The use of 24 kb/s coding for data transmission is not recommended.

Tests conducted indicate, Fig. 7, that for voice the 40 kb/s ADPCM coding performs approximately as well as 64 kb/s PCM according to Rec. G.711. Voice band data at speeds up to 12000 bits/s can be accommodated by 40 kb/s ADPCM. The performance of V.33 modems operating at 14400 bit/s over 40 kb/s ADPCM is for further study.

Under normal DCME operating conditions, no significant problems with DTMF signalling or with Group 2 and 3 facsimile apparatus are expected [8].

There are three modes of DCM operation so far identified
- Point-to-point mode
- Multi-Clique Mode (based on a limited multidestinational capability, with perhaps fixed but relatively small bearer capacities)
- Full Multi-Point Mode (based on fully-variable capacity allocation of

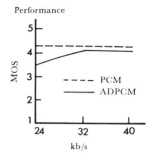

Performance

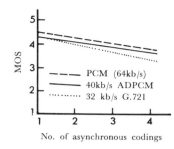

Fig.7. Subjective Speech Performance of Variable rate ADPCM Coding [8]

multi-destinational bearer channels).

A review of the activities of various CCITT study groups and of national bodies shows that current plans provide the means within DCME to accommodate the bearer services defined in Rec. I.211 Red Book sections 2.1.1 64-kb/s unrestricted, 2.1.2 64-kb/s useable for speech, 2.1.3 64-kb/s usable for 3.1 kHz audio, and 2.1.4 alternate speech/64-kb/s non-speech.

There are several issues concerned with DCM implementation which are being addressed as Question 31/XVIII by CCITT Working Party XVIII/8.

3.4 Other CCITT Activities for Future Standards

CCITT and other organisations (CEPT, Intelsat, Inmarsat etc.) have established a number of network applications which require speech bit rates less than 32 kb/s. As has been pointed out elsewhere in this book, 16 kb/s is the lowest bit rate today giving high quality of speech although coders operating at lower speeds exists which give adequate quality for applications in a circuit-oriented network environment or in packet networks.

The following main applications for the 16 kb/s speech coding have been identified by CCITT Working Party XVIII/8 Question 27/XVIII:
   i) Land Digital Mobile Radio (DMR) system and portable telephone;
   ii) Low C/N digital satellite systems. This include maritime thin-route and single channel per carrier satellite systems;
   iii) DCME. In this equipment low bit rate encoding is generally combined with DSI. The equipment may be used for long terrestrial connections and for digital satellite links generally characterized by high C/N ratios;
   iv) PSTN. This application covers the encoding of voice telephone

> channels in trunk, junction or distribution network;
>
> v) ISDN. This application is similar to that foreseen in PSTN, being understood that in this case end-to-end digital connectivity at 64 or 128 kb/s is available for multimedia applications such as video telephones (eg., 16 kb/s voice and 48 kb/s video),
>
> vi) Digital leased lines. Two possibilities may be envisaged in this case; one is where the end-to-end digital leased circuits include only one encoding/decoding, the other is where the end-to-end digital leased circuits are connected into the public network and they may include digital transcodings;
>
> vii) Store and Forward systems;
>
> viii) Voice messages for recorded announcements.

It has been agreed that CCITT should play the role of overall coordinator of activities related to the above applications in order to assist the various organizations in their studies in areas of common interest. This would allow the achievement of consistency between the performance requirements of the specific application and that of the overall network. This coordinating role is especially required in the definition of sensitive networking topics such as speech quality objectives, capability in terms of cascade transcoding and processing delay. The issue of CCITT guidelines on the forementioned topics would help to ensure international network interconnections with satisfactory overall performances.

The organisation involved in the early identification of the speech coding algorithms for specific applications have been invited to provide CCITT with punctual informations on networking issues and in particular to indicate their speech quality objectives in terms currently used in CCITT (eg. qdu and/or MOS or more discriminating tests).

The network performance parameters collected to date for the various applications are summarized in table II.

It is important to note that there are different priorities attached to the different applications and that if the urgent requirements are not tackled in a timely manner then there would be the possibility of increasing proliferation of 16 kb/s speech coding standards tied to Specific applications. Urgent action is required for applications (i) and (ii) in the table.

The European Telecommunications Administrations are in the process of planning a common digital mobile radio system which will be launched in 1991/1992. CEPT Working Group GSM (Group Special Mobile) which was set up in 1982 to coordinate Studies and activities (covering aspects of speech quality,

TABLE II
Network Performance Requirements for 16 kb/s Speech Coding

Network Requirements / Applications	Robustness against transmission errors	One way code 2/ decode 2 delay	Transcoding to existing standards / No of transcoding in cascade	Voiceband data / Other non voice services	Speech quality objectives	Possibility to operate at different bit rates	Sampling frequency (kHz)
(i) Land DMR systems and portable telephone	Acceptable quality up to 10^{-2} random errors Quality for burst error is under study	65 ms	to 64 kb/s / 2 asynchronous	No explicit requirements but tests will be performed / single and information tones	Comparable to that of 900 MHz analogue system	(-)	8
(ii) Low C/N digital satellite systems	No significant degradation with 10^{-3} random errors	60 to 80 ms	to 64 kb/s / 2 asynchronous	up to 2400 bits / single and DTMF tones	Comparable to that of companded FM (6 bit PCM)	needed	8
(iii) DCME	up to 10 random	40 to 80 ms	to 64 kb/s / 2 asynchr.	Yes / Tones	6 to 7 bit PCM	needed	8
(iv) PSTN	No significant degradation with 10^{-4} BER				6 to 7 bit PCM	not needed	8
(v) ISDN	as PSTN	(-)	to 64 kb/s / 4 synchr.	not needed	6 to 7 bit PCM	not needed	8
(vi) Digital leased lines	up to 10^{-4} random	70 ms	to 64 kb/s / (-)	(-) / tones	7 bit PCM	not needed	8
(vii) Store and forward systems	as PSTN	(-)	to 64 kb/s / (-)	not needed	6 to 7 bit PCM	(-)	8
(viii) Voice messages for recorded announcements	as PSTN	(-)	to 64 kb/s / (-)	not needed	less stringent than PSTN although speech intelligibility is required	(-)	8

(-) not assessed yet.

transmission delay, and complexity) has recenly selected [10] a speech coding algorithm which is of the linear predictive coding type at 13 kb/s rate using Regular Pulse Excitation and Long-Term Prediction: LPC (RPE-LTP).

INMARSAT is planning to introduce as new maritime satellite communication system from 1991 onwards which will provide users with high quality communication links even under adverse propogation conditions. INMARSAT is proposing a 16 kb/s Adaptive Predictive Coding (APC) algorithm which will meet the requirements shown in the table II.

In addition to the various 16 kb/s codec applications which require urgent CCITT action, several opinions have been expressed that CCITT should also undertake early activities on speech coding at around 8 kb/s in order to anticipate the likely development of autonomous standards in the near future such as the use of 8-9.6 kb/s for speech coding in DMR in order to effect better spectrum utilization.

The CCITT has just set up an Expert Group to establish whether it is possible to select a unique coding algorithm approach that meets requirements of the various network applications. Activities in this direction will likely develop in the next two years with the aim of minimizing the number of alternative coding techniques to be chosen as CCITT standards in the Study Period (1988-1992).

CCITT has also initiated studies for the Study Period 1988-92 regarding "Speech Packetization", "Encoding for stored digitized voice", and " speech analysis/synthesis techniques".

Packetized speech may find applications both for shortterm implementations, such as DCME [11] and for longer term applications, i.e., in the evolving broad-band ISDN when the "asynchronous transfer mode" (ATM) of operation will be implemented [8,12]. DCM applications are related to the use of digital links at speeds on the order of few Mbits/s, while ISDN-ATM applications are foreseen at much higher link speeds (i.e.,50-150 Mbits/s).

Among the problems to be studied the following items may be mentioned:

- Interfaces (1536/1984 kb/s)
- Speech coding algorithms (PCM, ADPCM)
- Voice-band data
- Error detection
- Voice delay
- Performance (packet loss and bit dropping).

The CCITT work (Question 29/XVIII) on "Encoding for stored digitized voice" assumes that the transmission of voice message among store-and-forward systems is in line with the message handling system (MHS) procedures specified in CCITT Rec. X.400 to X.420. It is also accepted that algorithms developed under "16 kb/s speech coding" could be used even for the encoding of the stored voice, especially when associated with a suitable silence encoding.

The general requirements for the encoding of stored voice are tentatively given by CCITT as follows:

- low bit rate possibly using silence coding
- high quality speech (equivalent to 6 to 7 bit PCM)
- speaker recognizability
- variable rate operation, i.e. graceful degredation of voice quality when the bit rate is decreased
- robustness in multi-speaker conditions and with typical basic ground office noise.

Standards for voice storage services are likely to cover bit rates from 4 to 16 kb/s. The bandwidth is likely to be about 3 kHz, but it is too early for CCITT to settle on a coding technique. It is to be noted that coding delay will be much less of a problem here than say in the packetised speech with real-time conversations.

As far as question 32/XVIII on "speech analysis/synthesis techniques" is concerned there has not been much activity within CCITT Working Party XVIII/8 even though many member countries have been very active in this field with encouraging results. The only contribution reaching CCITT seems to have come from INMARSAT who reported on activities undertaken outside CCITT to proceed towards 4.8/9.6 kb/s encoding standards by the AEEC (Airline Electronic Engineering Committee) for telephony applications from commercial airplanes.

4. NATO STANDARDISATION ACTIVITIES IN SPEECH PROCESSING

Like the National Security Agency in the USA and similar agencies in the other countries, NATO also has not waited for international agreements and has set standards for voice coding at rates ranging from 2.4 kb/s to 16 kb/s. An overview of some of these standards is given below.

4.1. NATO STANAG 4198

The NATO standardisation Agreement (STANAG) 4198 which was promulgated on 13 February 1984 by the NATO Military Agency Standardisation (MAS) defines the voice digitizer characteristics, the coding tables and the bit format requirements to ensure the compatibility of digital voice produced using 2400 b/s Linear Predictive Coding (LPC).

The content of this agreement is outlined below, as an indication of what needs to be specified in order to assure interoperability between equipments manufactured by different nations.

a) Description of Linear Predictive Coding

Figs. 8 and 9 give the block diagrams of the transmitter and receiver portions of a typical LPC system.

i) The input bandwidth must be as wide as possible, consistent with a sampling rate of 8 kHz. It is desirable that the pass band be flat within 3 dB from 100 to 3600 Hz.

ii) After first order pre-emphasis $(1-0.9375 Z^{-1})$ 10 predictor coefficients are determined by linear predictive analysis.

iii) For pitch and voicing analysis, 60 pitch values are calculated over the frequency range of 50 to 400 Hz. A two-state voicing decision is made twice per 22.5 milliseconds frame.

iv) The excitation and spectrum parameters are then coded and error corrected for transmission at 2400 b/s.

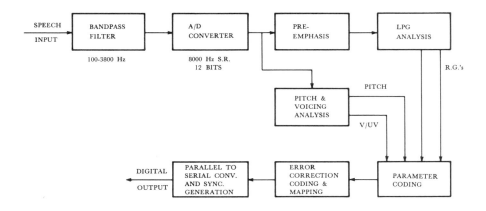

Fig.8. Linear Predictive Coder Transmitter

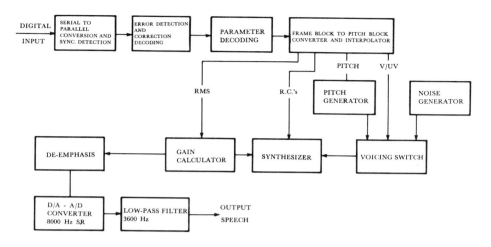

Fig.9. Linear Predictive Coder Synthesizer

b) Voice Digitizer Characteristics.

Sampling Rate	8 kHz ± .1%
Predictor Order	10
Transmission Data Rate	2400 b/s ± .01%
Frame Length	22.5 ms (54 bits per frame)

Excitation Analysis

Pitch	50-400 Hz, semi-logarithmic coding (60 values)
Voicing	A two-state voicing decision is made twice a frame
Amplitude	Speech root-mean-square (rms) value, semi-logarithmic coding (32 values)

Spectrum Analysis

Pre-emphasis	Typical first order digital transfer function $1 - .9375Z^{-1}$
Spectrum Approximation	10th order all-pole filter
Spectrum Coding	Log area ratio for the first two

coefficients and linear reflection
coefficients for the remainder

Transmission Data Format

Synchronization	1 bit
Pitch/Voicing	7 bits
Amplitude	5 bits
Reflection Coefficients	41 bits for 10 coefficients if voiced, or 20 bits for 4 coefficients with 20 error protection bits if unvoiced

Error Detection and Correction

Voicing Decision	(1) Full-frame unvoiced decision encoded as a 7-bit word having a Hamming weight of zero (7 zeros)
	(2) Half-frame voicing transition encoded as a 7-bit word having a Hamming weight of seven (7 ones)
Unvoiced Frame Parameters	Hamming [8,4] codes to protect most significant bits of amplitude information and first 4 reflection coefficients.
Voiced Frame Parameters	(1) 60 pitch values mapped into 60 of 70 possible 7-bit words having a Hamming weight of 3 or 4
	(2) Typically for good performance under error conditions an adaptive smoothing algorithm should be applied to pitch, amplitude and first 4 reflection coefficients for eradication of gross errors based on the respective parameter values over three consecutive frames

Synthesis

The synthesis filter must be a 10th order all-pole filter with appropriate excitation signals for voiced and unvoiced sounds capable of satisfying the speech intelligibility requirements as specified in Section (d) below.

The typical de-emphasis transfer function is $1 / (1 - 0.75\ Z^{-1})$

A recommended 40 sample all-pass excitation for voiced speech is as specified in the STANAG.

c) Interoperable Coding and Decoding

The RMS, reflection coefficients, pitch and voicing are coded to 2400 b/s. The frame length is 22.5 ms. The bit allocation for the voiced and non-voiced frames, the specified transmitted bit stream for voiced and non-voiced frames, synchronization pattern, coding of the reflection coefficients and the logarithmic coding of RMS have to be specified as in the tables given in the Stanag.

d) Performance Characteristics

i) The performance of LPC-10 speech processors shall be measured in terms of intelligibility test and free conversation test [14].
ii) The voice intelligibility of the voice processor shall be measured using the Diagnostic Rhyme Test (DRT-IV).For the DRT, English,American and French versions are to be used and the talkers and listeners are to be familiar with the language in each case. The input analogue tapes to be used for the English,American DRT and the minimum acceptable scores, which should be obtained from an independent contractor, are given overleaf.
iii) The Free Conversation Test shall be carried out using at least 6 pairs of subjects who shall have no undue difficulty in conversing over a normal telephone circuit. A mean opinion score of at least 2.5 shall be obtained when the speech is transmitted between typical office environments and with zero bit error rate.

Acoustic Environment	Talkers	Tapes	Bit Error Rate	Microphone	Minimum Acceptable Score
Quiet	6M	E-1-A E-1-B	0	Dynamic	86
Office	3M	C-4-A	0	Dynamic	84
Shipboard (Saipan)	3M	K6-1.2-A	0	H250	85
Aircraft (P-3C)	3M	K7-1.2-A	0	EV985	82
Jeep	3M	K8-1.2-A	0	H250	82
Tank	3M	K9-1.2-A	0	EV985	82
Quiet	3M	E-1-A	2.0%	Dynamic	82
E3A	3M	K1-11A	0	215-330	82
F15	3M	K-10-1	0	M101	75
F16	3M	1C-11-1	0	M101	75
TORNADO F-Z					

Note : The above performance tests have recently been changed by a NATO Working Group on narrow-band systems; the type of test to be used is not any more dictated and may be any suitable test (see Chapter 6), where the existing LPC-10 algorithm is used as reference together with the defined noise conditions.

e) STANAG's Related to STANAG 4198

There are NATO STANAG's which specify the modulation and coding characteristics that must be common to assure interoperability of 2400 b/s linear predictive encoded digital speech transmitted over HF radio facilities (STANAG 4197, promulgated on 2 April 1984), and on 4-wire and 2-wire circuits (STANAG 4291, promulgated on 21 August 1986).

4.2. 4800 b/s Voice Coding Standard

There is a US Proposed Federal-Standard (PFS-1016), to be considered also by the US Military and NATO, for a 4800 b/s voice coder which is claimed to outperform all US government standard coders operating at rates below 16 kb/s and even to have comparable performance to 32 kb/s CVSD and to be robust in acoustic noise, channel errors, and tandem coding conditions [15].

PFS-1016 is embedded in the US proposed Land Mobile Radio standard (PFS- 1024) that include signalling and forward error correction to form an 8 kb/s system. A real-time implementation of a 6400 b/s system with an embedded PFS-1016 is said to be submitted for consideration in INMARSAT's standard system.

The coder, jointly developed by the US DOD and ATT Bell Laboratories, uses a code excited predictive (CELP) coding which is a frame-oriented technique that breaks a sampled input signal into blocks of samples (i.e., vectors) which are processed as one unit. CELP is based on analysis-by-synthesis search procedures, two-stage perceptually weighted vector quantization (VQ) and linear prediction. A 10th order linear prediction filter is used to model the speech signal's short-term spectrum and is commonly referred to as a spectrum predictor-long-term signal periodicity is modeled by an adaptive code book VQ (also called pitch VQ because it often follows the speaker's pitch in voiced speech). The residual from the spectrum prediction and pitch VQ is vector quantized using a fixed stochastic code book. The optimal scaled excitation vectors from the adaptive and stochastic code books are selected by minimizing a time varying, perceptually weighted distortion measure. The perceptual weighting function improves subjective speech quality by exploiting masking properties of human hearing.

4.3. NATO Draft STANAG 4380

The draft STANAG 4380 has been prepared by the "Subgroup on Tactical Communications Equipment" of the "Tri-Service Group on Communications and Electronic Equipment" (TSGEE) and has been forwarded to the Major NATO Commanders for review/comment and to the Nations for ratification. The STANAG deals with Technical Standards for Analogue-Digital Conversion of Voice Signals using 16 kb/s delta modulation and syllabic companding controlled by 3-bit logic (CVSD). A block diagram of the coder/decoder is shown in Fig 10.

The following information is given as an indication of the standards that are required to ensure interoperability, where and when required, of 16 kb/s digital voice signals for tactical communications.

184

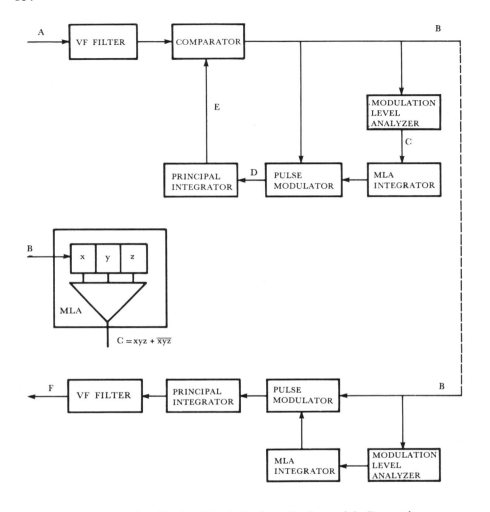

Fig.10. 16 kb/s Delta Modulation Codec with Symatic
Companding (CVSD)

a) Frequency Response
The input and output filters shall have a passband of at least 300 Hz to
2.7 kHz.

b) Modulation level
When an 800 Hz sinewave signal at 0 dBmO is applied to the input of the
coder (point A in Fig 10), the duty cycle at the output of the modulation
level analyser(point C) shall be 0.5 (The duty cycle is the mean proportion

of binary digits at point C, each one indicating a run of three consecutive bits of the same polarity at point B).

c) Companding

In both the coder and the decoder the maximum quantising step, which drives the principal integrator at point D, shall have an essentially linear relationship to the duty cycle;the ratio of maximum to minimum quantising steps at the decoder output (point F) shall be 34 dB±2 dB.

With the decoder output (point B) connected to the decoder input (point B'), when an 800 Hz sinewave at the coder input (point A) is changed suddenly from -42 dBmO to 0 dBmO, the decoder output signal (point F) shall reach its final value within 2 to 4 ms.

d) Distortion and Noise

When a sinewave signal at -20 dBm0 is applied to the coder input (A), the attenuation distortion at the decoder output(F), relative to that at 800 Hz, shall be within the limits that are specified in the STANAG; the distortion contributed by the coder alone, measured at the output of the principal integrator (E), is also specified. The idle channel noise at the output of the decoder (F) shall not exceed - 45 dBmO and the level of any single frequency in the range 300 Hz - 8 kHz shall not exceed -50 dBmO.

The limits for the signal/noise ratio at the output of the decoder (F) are also given in the draft STANAG 4380.

5. CONCLUSIONS

A quantitative description of the current state of telephone speech coding in terms of standarts activity, bit rate, and MOS is summarized in Fig. 11. The solid curves in this figure refer to generic examples of coding algorithms outlined in the text above and the broken curve represents a research goal which is regarded as achievable. The solid dots refer to coding algoritms that provide high quality at 64,32 and 16 kb/s.

In the past, standards used to follow technology. Manufacturers dominated standards activities. Contentions were avoided by adoption of multiple options in the standards. Interoperability of products was not an issue. The marketplace was not sensitive to the speed of standards processes.

Today, users have a great influence in standards development. They demand interoperability of products, and they reject nonresolution of contentions.

186

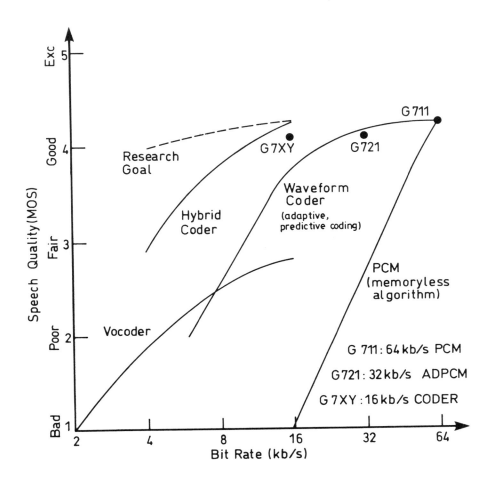

Fig.11. 3.2 kHz Speech Quality and Transmission Rate [17]

Standards processes are protracted and, in sharp contrast with the past, standards now lead technology.

Standards affect customers, service providers, equipment manufacturers, and vendors and the growing importance of standards is now becoming universally recognized. In the future, this importance will be driven even higher by the increasing complexity of the technology and the rising expectations of the world's growing population. And, as a result of this recognition, a number of major trends are developing within the telecommunication industry. These are:

    * Consumers and users in the industry are increasingly demanding compliance with standards.

    * Standards activities are growing at a rapid rate.

    * There is a growing sense of urgency in standards.

These major trends can only be satisfied through increasing the responsiveness by standards bodies. Standards bodies must be sensitive and responsive to the growing needs for both timely delivery of accepted standards and conservation of global resources in doing standards work.

Today, standards-making organizations are mainly bottom-up driven. Direction is charted by the individual contributions submitted by members. Studies are started by proposals from members, they are supported by member contributions and they are completed when contributions cease and a consensus is reached between the members. This approach is acceptable and necessary but should be complemented by some top-down drive for timeliness and responsiveness to whatever the requirements are.

6. REFERENCES

[1] Bellchambers W.H. et al., "The International Telecommunication Union and Development of Worldwide Telecommunications", IEEE Com. Magazine, Vol. 22, No.5, May 1984.

[2] "Digital Networks, Transmission Systems and Multiplexing Equipment" CCITT Red Book, Vol III-Fascicle III-3, Geneva 1985.

[3] Decina, M. and Modera G., " CCITT Standards on Digital Signal Processing" IEEE Selected Areas in Communications Vol.6, No.2, Feb., 1988.

[4] CCITT SG XVIII, Rep.R.26(C), Working Party 8, Geneva Meeting, Aug., 1986.

[5] Benvenito N.et al, "The 32 kb/s ADPCM Coding Standard", ATT Tech.J., Vol.65, Sept/Oct. 1986.

[6] CCITT Report COM XVIII - R26(C) part C.2, "Draft Recommendation G72X: 7kHz Audio Coding within 64 kb/s", July 1986.

[7] Maitre X.,"7 kHz Audio Coding within 64 kb/s",IEEE Journal on Selected Areas in Communications, Vol.6, No.2, Feb. 1988.

188

[8] CCITT SG XVIII, Rep. R45(C), Working Party 8, Hamburg Meeting, July, 1987.

[9] Gerhauser, H.L.,"Digital Speech Interpolation with Predicted Wordlength Assignment", IEEE Trans. on Coms., Vol. COM-30, No.4, April 1982.

[10] Natvig, J.E., "Evaluation of Six Medium Bit-Rate Coders for the Pan-European Digital Mobile Radio System", IEEE Journal on Selected Areas in Communications, Vol.6, No.2, February 1988.

[11] "Packet Speech and Video", IEEE Journal on Selected Areas in Coms., Vol.7, No.5, June 1989.

[12] "Broadband Packet Communications", IEEE Journal on Selected Areas in Coms., Vol.6, No.9, December 1988.

[13] Muise, R.W. et al., "Experiments in Wideband Packet Technology" International Zurich Seminar (IZS'86), March 1986.

[14] NATO Documents: AC/320-D/159:AV/302(NBDS)D/13, dated 11 June 1981.

[15] Campbell, J.et al., "An Expandable Error-Protected 4800 b/s CELP Coder" Proc.ICASSP, 1989.

[16] Mermelstein, P., "G-72z, A New CCITT Coding Standard for Digital Transmission of Wideband Audio Signals", IEEE Com.Mag., Jan. 1988.

[17] Jayant, N.S., "High Quality Coding of Telephone Speech and Wideband Audio",

CHAPTER 8

APPLICATION OF AUDIO/SPEECH RECOGNITION FOR MILITARY REQUIREMENTS

EDWARD J. CUPPLES
and
BRUNO BEEK

ROME LABORATORY
GRIFFISS AIR FORCE BASE NY 13441-5700

INTRODUCTION

Increases in the functional capabilities of military systems have made these systems increasingly more difficult to operate. Increased operator workload in modern workstations and aircraft have produced operator stress and fatigue, resulting in degraded operator performance, especially in time critical tasks. One reason for this problem is that both data entry and system control functions are often controlled via the systems keyboard. In some systems functions are nested many layers deep making the system inefficient and difficult to use. For this reason technology to improve the interface between the system and its operators is of high interest. Many efforts and several technologies are being pursued in speech recognition and synthesis, multimodal interface techniques, and voice interactive concepts and methods. Such work is being pursued to satisfy the requirements for modern communication, collection, analysis, identification, resource management, and control.

Interest in potential uses of Automatic Speech Recognition (ASR) technology is steadily increasing in both military and civilian communities. Much of this interest is due to advances in electronics and computers rather than in new techniques for speech recognition. Despite its current limitations, ASR promises to aid in a variety of military applications by increasing the effectiveness and efficiency of the man machine interface. Indeed, military organizations have long been, and continue to be, one of the main sources of support of research and development of ASR technology.

This chapter discusses some recent applications of ASR technology. It is not intended to be exhaustive but rather presents a representative perspective of the military uses of this technology. Four major categories of applications are discussed: Audio Signal Analysis, Voice Input for Command and Control, Message Sorting by Voice, and Automatic Gisting.

AUDIO SIGNAL ANALYSIS

Speech enhancement and interference reduction technology to improve the quality, readability and intelligibility of speech signals that are masked and interfered with by communication channel noise has high interest and many applications. This interest in speech enhancement is not only in improving the quality, readability and intelligibility of speech signals for human listening and understanding but to improve speech signals for machine processing as well. Speech technology such as speaker identification, language recognition, narrowband communications, and word recognition being developed requires good quality signals in order to provide effective results. The development of automatic real-time speech enhancement technology is therefore of very high interest to military users.

There are a large number of applications for speech enhancement. Many systems that perform silence or gap removal and/or speech compression have difficulty with the processing of noisy communications data. In many instances gap removal is completely ineffective and compression schemes completely degrade speaker identity and cause large reductions in intelligibility. These systems require speech enhancement to be operationally effective. The use of Automatic Speech Recognition (ASR) in noisy environments such as the Cockpit is of very high interest. Al though there has been some success in using restricted and well structured ASR in the cockpit, difficulties with acoustic noise in the airborne environment is much more troublesome for larger vocabulary continuous speech recognition systems. The successful use of enhancement for ASR can offer performance improvements that will make voice control and data entry operationally acceptable for many airborne applications.

Another area in which the noise generated in an aircraft causes problems is the use of vocoder for narrowband jam resistant communications. Vocoder technology use is restricted in many airborne applications because the acoustic noise generated by the aircraft degrades the intelligibility of the vocoder system to an unacceptable level. Speech enhancement to reduce the aircraft noise offers the capability to make a variety of vocoder technology available for airborne use.

Speech Enhancement

Speech Enhancement is the capability to remove frequently encountered communication channel interferences with minimum degradation to the speech signals. The types of interferences or noises removed can be classed into three groups (1) impulse noise such as static and ignition noise, (2) narrowband noise which includes all tone-like noise, and (3) wideband random noise such as atmospheric, receiver electronic noises, and aircraft noise. Impulse noise removal processes are usually a time domain process. The process is very effective for removing impulses up to 20 milliseconds in length. Narrowband noise removal processes are usually a frequency domain process. Techniques are required to remove both high level and low level tones of which there may be several hundred. The tones may be fixed or moving. Such a capability is extremely useful in removing power converter hums and hetrodyne signals found communication channels. Wideband noise removal processes frequently use cepstrum subtraction processes. The only successful process developed by the United States Air Force at Rome Laboratory, Griffiss AFB NY is a subtractive process that is accomplished in the spectrum of the square root of the amplitude spectrum. While this function is not the same as the cepstrum (the cepstrum is the spectrum of the log amplitude spectrum), since it resembles the cepstrum it is referred to as the root-cepstrum. In this method of noise reduction the average root-cepstrum of the noise in the input signal is continually updated and subtracted from the root-cepstrum of the combined speech and noise. Because the random noise concentrates disproportionately more power in the low region of the root-cepstrum than does the speech, the subtracted reconstructed time signal produces an enhanced speech signal. A picture of a prototype, the VLSI, and the VHSIC enhancement units is shown in Figure 1.

Other enhancement processes have been developed for unique interferences and noises and have been shown to be effective for their specific use. New enhancement methods are continually sought. Some examples currently being pursued by research and development laboratories are technologies such as Hidden Markov Models, Neural Networks, and Artificial Intelligence.

Speech Enhancement Testing

How to determine the value of a speech enhancement capability or technique has been a never ending debate. Test methods are subjective and test procedures not well established making comparisons of systems and techniques difficult to impossible. However, proper testing is critical to successfully applying and fielding a capability in a military or commercial application. As a result, several test methods and procedures were developed by Rome Laboratory through necessity. Since the Rome Laboratory Speech Enhancement Unit (SEU) is currently the only process that has been thoroughly tested in the laboratory and in field military applications, and since many of the test methods have been endorsed by other agencies, the discussion on testing and on enhancement capabilities will use the SEU as an example.

The SEU has been tested in two areas. They are (1) the reduction of communication channel noise to improve the recognition performance of human listener and (2) the reduction of wideband random noise and aircraft cockpit noise to improve the performance of automatic speech recognition (ASR) systems. Improvements in performance have been demonstrated in other areas.

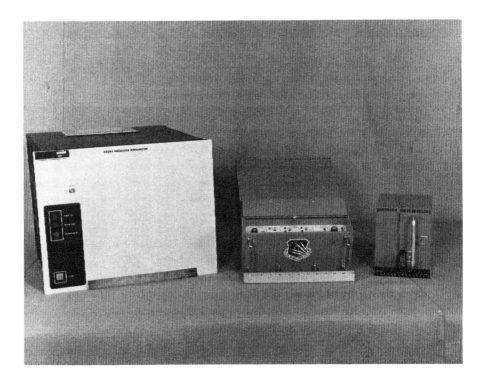

Figure 1. Evolution of Speech Enhancement Hardware

From left to right: Prototype, VLSI, VHSIC

The first test conducted on the SEU determined the effect of processing radio frequency voice communication channels containing a variety of off-the-air noises on the monitoring performance of humans. The signals were monitored by equally skilled trained Air Force operators both before and after enhancement by the SEU. The data was controlled so that no operator heard the same data before and after enhancement. The readability of the signals was rated before and after enhancement. The readability of the signals is as shown in Figure 2. Note the shift in the readability of the signals after enhancement. The results clearly show an improvement in readability. However, not only was there a significant improvement in the readability of the signals but operator fatigue was reduced, intelligibility improved, and very importantly, the enhancement process was found to be capable of being operated in an entirely automatic mode. Also important was the uncovering of events that were not recognized before enhancement. These results appear to agree with equipment laboratory tests which showed the narrowband and impulse noise to be attenuated as much as 40 dB.

Measurements on the wideband removal process showed a signal-to-noise ratio improvement of from 15 to 21 dB.

The second set of tests were conducted to determine the effect of using the SEU as a preprocessor to automatic speech recognition systems. Several speech recognizers were used with good results.

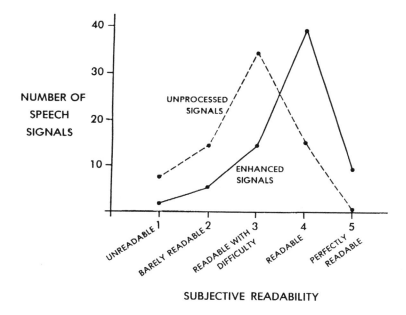

Figure 2. Speech Enhancement Operational Results

The results of a test conducted at an Air Force flight laboratory with the SEU acting as a preprocessor to an LPC-based recognizer showed substantial recognition improvements. The tests were conducted in a facility where the acoustic environment of the F-16 cockpit was simulated. The tests were conducted using the Advanced Fighter Technology Integration (AFTI) 36-word vocabulary. Training was accomplished without the SEU and in 85 dBa sound pressure level (SPL). Six subjects were tested; four military pilots and two Lear Siegler personnel. The two subjects used for the enhancement tests were the lowest scoring military pilots in the tests. Enhancement was used only during the 109 dBa and 115 dBa noise level tests. At 109 dBa noise level recognition performance increased from 46% without SEU processing to 75% after enhancement. Performance jumped from 30% to 79% after enhancement for the 115 dBa noise level condition, Figure 3.

Other tests using the SEU or a preprocessor have shown varying degrees of improvement. Test results without training the recognizer through the SEU show digit recognition improvements of 20% correct recognition to 83% after enhancement for an input S/N of 3 dB using wideband random noise, Figure 4.

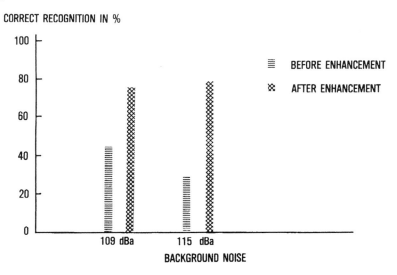

Figure 3. SEU/LPC Based Recognition Performance

Better performance was obtained by training the recognizer through the SEU under no noise conditions. The results obtained for this condition show an improvement from 21% to 100% correct recognition after enhancement at a 10 dB S/N.

Co-Channel Interference Reduction

Another type of interference that is often encountered in military operations is co-channel interference. This voice on voice interference occurs when there are two or more transceivers transmitting at the same time on the same frequency channel. As an example, co-channel interference occurs when two aircraft attempt to communicate with the control tower at the same instant in time which most often results in unintelligible speech at the tower.

Techniques to reduce co-channel interference have been applied to the radio frequency, the intermediate frequency and the demodulated audio signals. Radio frequency methods attempt to reduce the interference by creating antenna peaks and nulls to separate signals by spatial filtering. This method is limited by the physical size

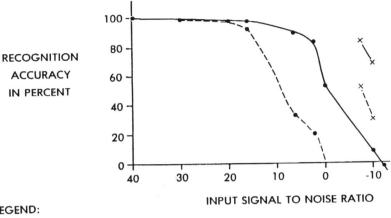

RECOGNITION
ACCURACY
IN PERCENT

INPUT SIGNAL TO NOISE RATIO

LEGEND:
- - - WITHOUT ENHANCEMENT
—— ENHANCED
• WIDEBAND NOISE
× TONAL NOISE

Figure 4. Recognition Performance of Filter

Bank Word Recognizer

of the antenna array. This method is not effective for signals having the same line of bearing.

Intermediate frequency techniques which have had some success use signal amplitude and frequency to discriminate between the co-channel signals. The techniques are effective for separating frequency modulated (FM) signals from other FM signals and amplitude modulated signals (AM) from FM signals but are not effective for separating AM from AM signals.

Separating the co-channel signals in the audio band is attractive because it is independent of the geometry of the emitters and the signal modulation. Many techniques have been developed but generally all these techniques are based on separating voiced speech segments only. Therefore these techniques must rely on separating the pitch of the talkers and using this information to extract the desired talker or suppress the interfering talker. Some of the techniques that have had some success on signals where the ratio of the talkers strengths are large (-12 to -25 dB reference to the stronger talker) are pitch tracking comb-filters and harmonic magnitude suppression algorithms. Both require accurate pitch tracking and hence have difficulty where the pitches of the talkers cross, where the pitch changes rapidly and where the pitches are close in frequency. Analysis techniques that provide high resolution to resolve small differences in the pitch of the talkers are not able to track rapid changes in pitch.

Therefore trade-offs are generally made between frequency and time resolution to satisfy both tracking speed and frequency resolution requirements.

Techniques that rely on eliminating the pitch components of one talker by removing the components entirely have not had success in improving the intelligibility of the desired talker. The suspected reason is that some components of the desired talker are also eliminated and the holes in the frequency spectrum generate masking noise. The only techniques that have improved intelligibility of co-channel speech are those that have suppressed the larger interfering talker to a level near to the desired talker. This places the burden of separating the talkers on the listener. Results for a method called Harmonic Magnitude Suppression (HMS) which uses this technique show an absolute improvement in intelligibility of 6.3% for a ratio of -6 dB reference stronger talker and 7.2% for a ratio of -18 dB for the same reference. It should be remembered that this improvement was made by processing only the voiced speech. Unvoiced speech was left for the listener to interpret. Based on the continuous speech of two English talkers simultaneously talking, the percent of the time voiced speech falls on voiced speech is approximately 35%. Hence, if the HMS process provided perfect separation of voiced-on-voiced speech, the maximum improvement in intelligibility can be no greater than 35%. Larger improvements will require research in the separation of unvoiced speech from both the voiced and unvoiced speech of another talker.

VOICE INPUT FOR COMMAND AND CONTROL

There are many man-machine interface (MMI) problems associated with the modern communication stations, battle management workstations and the advanced aircraft cockpit. Several factors have led to the MMI problems and the subsequent thrust of military development of MMI technologies. They are:

o Adding on new capabilities to existing systems

o New systems with many combined capabilities

o Increased complexity of the environment

o Reduced time to complete tasks

o Increases in the number of time critical tasks

Many of these factors are the direct result of reduced manpower (accomplish more with fewer operators), and the increased speed of events caused by higher speed aircraft and advanced weaponry.

New ASR and speech synthesis technology forms the basis for voice input/output (I/O) systems. Such systems can improve man-machine interaction or modern communications, collection, analysis, identification, resource management and control. Speech communication with machines can offer advantages over other modes of communication such as manual methods, especially when humans are engaged in tasks requiring hands and eyes to be busy. Speech offers the most natural, and potentially the most accurate and fastest mode of communication, but is susceptible to environmental interference, and restricted by speaker and training requirements. Researchers are currently investigating speech recognition techniques which would permit a more natural, continuous form of speaking style and which would require a minimum amount of training by the speaker.

The workload of the military flight crew is becoming more demanding, due to increases in the amount of complex equipment crew members must monitor and control. Hence, there are constant demands on crew members for manual, visual and aural attention in order to perform vital mission functions, such as navigation, controlling weapons and monitoring sensors. At present, most critical functions are performed via manual operation of switches and keys. The increase in the number of manual tasks, as well as information processing demands, has made it difficult for the crew member to perform all the necessary functions while maintaining control of the aircraft. ASR technology can aid in relieving this information and motor overload by allowing the use of voice to control manual functions.

An airborne environment presents serious problems for any speech recognition device. These problems include high ambient noise, high g-forces, vibration, effects of oxygen masks, and extremes of altitude, pressure, temperature and humidity. Yet there is no doubt that military organizations see ASR technology as an integral part of future airborne cockpit avionics if the challenge of operating in the harsh air environment can be met.

Currently efforts are being made for the development of MMI concepts and testbeds for test and evaluation of those concepts. The overall purpose of the research is to determine the requirements to provide efficient interfaces for the advanced cockpits and workstations. The voice interface goals are to develop the rudiments of an overall philosophy for verbal interaction with these systems.

In order to develop the philosophy and subsequent techniques, detailed scenarios for the cockpit and workstations are analyzed in terms of tasks, workload types, type and amount of information to be transferred, time constraints, criticality of the information, and environmental conditions. Using the scenarios information, experiments are conducted to determine fundamental relationships such as:

o   the effects of S/N in terms of time and accuracy on the completion of an audio task at various audio workloads.

o   the effects of various visual, manual, and oral workloads on various audio (listening) tasks and vice versa.

o   the effects of injecting audio messages (both voice and sound) into a system under various audio, visual, and manual workloads.

Knowing these interrelationships narrows the number of interface modalities for a given task under a given set of conditions and allows an estimation of a performance level. Based on the results from the experiments, designs for MMI testbeds are developed, and evaluated. Tests are generally conducted using the communications scenarios.

Voice Verification

A different type of voice command system is used to control entry to secure areas and computer systems. There is significant military interest in the use of automated systems based on personal attributes (such as speech) to verify the identity of individuals seeking access to restricted areas and systems (such as flight lines, weapon storage areas, classified record storage areas, command posts, computers, workstations, aircraft, etc.). In this application, ASR technology is employed for automatic speaker verification, which identifies who is doing the talking rather than the words being spoken. Techniques based on both amplitude spectral information and Linear Predictive Coding (LPC) have proved successful.

In discussing the accuracy of speaker verification and other ASR systems, it is important to note the tradeoffs that can be made which affect system's performance. The two most commonly recorded error types are: rejection (a legitimate utterance is falsely rejected) and substitution (an incorrect utterance of falsely substituted for the legitimate utterance). In evaluating speaker verification performance, rejections are called "Type I" errors and result when an authorized user has been incorrectly denied access to a secure area. Substitutions are called "Type II" errors and are a consequence of an imposter succeeding in gaining access as an authorized user. The tradeoff between the two error types are illustrated in Figure 5.

Most ASR systems (including speaker verification) incorporate a variable threshold which can be adjusted to control the balance between error types. Lowering the threshold tightens the requirements for acceptance of an utterance and thus lowers the Type II error, but with an increase in the Type I error. Also shown in Figure 5, by the dotted curve is a Receiver Operating Characteristic, which is a graph of the overall recognition accuracy as a function of threshold.

Recognition accuracy may be increased by threshold adjustment, with however, a penalty of additional substitution errors (Type II errors).

In one test of an automatic speaker verification system intended for military use, the average Type I and Type II error rates were both on the order of one percent. The test included over 100 talkers, over a several month test period (which included occasions when speakers had colds or other voice ailments), and for an environment with a high signal-to-noise ratio (SNR). This system was able to perform successfully even when several professional mimics attempted to imitate selected target speakers.

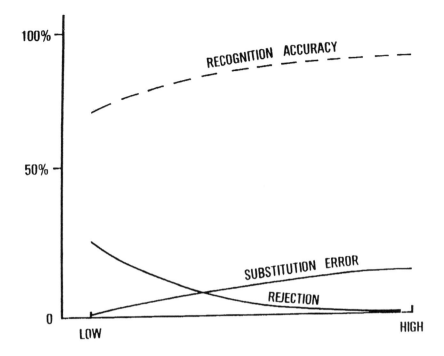

Figure 5. Speaker Verification Performance

Recent results obtained in a speaker verification test using 100 male and 100 female speakers, show a 1% Type I error for 7300 verification attempts and a 0.07% Type II error for 28000 verification attempts.

It is important when using ASR technology for military command and control applications that the total system be considered, not just the voice component. A thorough analysis of the human job tasks and a complete understanding of the system and environment to which ASR technology is interfaced are necessary.

MESSAGE SORTING/AUDIO MANIPULATION

Listening to radio broadcasts is a time-consuming, manpower-intensive and tedious task for military operators. This is due to the high density of received signals and the poor signal quality, which causes operator fatigue and reduced effectiveness. A potential solution to the problem is the use of ASR technology to automate part of the listening process. There are several recognition technologies being pursued that address the message sorting and routing problem, these include speaker identification, language recognition and keyword recognition.

Figure 6 shows how a typical message-sorting system might operate. A number of voice channels are multiplexed into a preprocessor. The preprocessor performs initial signal processing such as noise and interference removal as well as co-channel interference reduction. Subsequently, several automated ASR systems such as speaker authentication, language recognition, etc. determine important characteristics of the speech in the channels being monitored. Based on these results, a decision processor then comines the information from the ASR system with signal related inputs such as signal-to-noise ratio, time, etc., and provides a correlated input to a switching demultiplexer which provides the messages to an operator, or an automatic gisting system.

In order to satisfy military operational needs these recognition technologies must handle several operational constraints.

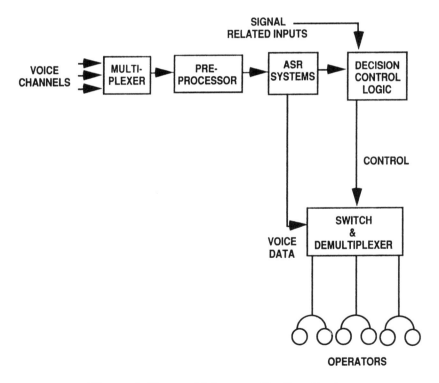

Figure 6. General Message Sorting System

An ASR system must

o Be context independent for speaker and language identification

o Handle uncooperative speakers

o Be robust to band-limited and noisy channels

o Handle dynamic channel conditions

o Operate on-line and in real-time

o Perform recognition on very short messages

Speaker Authentication

Speaker authentication is one method of message sorting that can be used to reduce the number of signals a communications operator must handle. Such systems must identify unknown talkers on multiple channels in real time using a small sample of their speech and under the above operational constraints. The operator can specify those talkers who are of interest at a particular time, and the system will route to the operator only speech that it identifies as spoken by the specified talkers.

Prior to executing a recognition task, a speaker authentication system is trained using one to two minutes of speech from each of the talkers who may later be recognized. The major requirement for the system is that it identify speakers using as little as two to five seconds of their speech since messages are often short but critical. Very few systems have been field tested.

One Speaker Authentication System developed by Rome Laboratory uses two techniques, a multiple parameter algorithm using the Mahalanobis metric and an identification technique based on a continuous speech recognition (CSR) algorithm. The multiple parameter algorithm uses both speech and non-speech frames. The speech frames are used to characterize the talker for recognition, and the non-speech frames to detect possible changes in talkers.

Recognition is performed by comparing the current average parameter vector with each of the active speaker models. Once per second the identity of the three models that are closest to the speech being recognized are output with their corresponding scores. Each second, the frames from the last second are accumulated and added to the average. The distance is then computed using the Mahalanobis metric.

The recognition module also monitors non-speech frames to detect pauses in the input speech that are associated with possible changes in talkers. When non-speech frames are input, the recognition module ignores the frame, but increments the silence-frames-in-a-row counter. If the silence-frames-in-a-row counter exceeds a silence threshold (user selectable, default value of 0.5 seconds) the recognition module signals a possible change in talker.

A second approach uses small sub-word templates to model a person's voice characteristics, rather than the long term spectral statistics that are used in the multi-parameter technique. The test results, using a CSR speech recognition system, show a very significant improvement in recognition accuracy over the first approach. The recognition accuracy exceeded 95% for clean speech segments of 2 seconds or longer duration, as compared to 75% for the multi-parameter technique. Speech enhancement as a noise removal preprocessor to speaker identification is required if automatic message sorting by speaker is to be effective in field operations.

Audio Management

Increases in the functional capabilities of modern workstations have made them increasingly more difficult to manage and operate. Increased operator workload has produced operator stress and fatigue, which has resulted in degraded operator performance, especially in time critical tasks. Military research and development continues to investigate and develop methods for audio handling, routing, and prioritization.

Rome Laboratory developed the first Advanced Speech Processing Station (ASPS) in the late 1970's. The concept of the ASPS was to alleviate the problems associated with analog recording methods by utilizing digital techniques. These techniques were the first to allow an operator to playback pre-recorded speech while still recording incoming speech. Utilizing a two minute buffer, digital techniques allowed the operator to manipulate the audio signal in the following ways: jump backwards or forwards, speed-up or slow-down while retaining frequency information, repeat or loop speech segments, tag speech for instant recall and remove silence or non-speech gaps.

New systems improved both the audio and text capabilities, provided better operator interfacing, and reduced workstation size, weight and cost. Tests on these systems have demonstrated improved performance/productivity (speed and accuracy), reduced operator fatigue and improved comprehension of the audio data.

Because of the success of these techniques, modern workstations containing many of the capabilities are commercially available. Audio manipulation capabilities are also available in a stand alone unit, Figures 7 and 8.

AUTOMATIC GISTING

There is high interest in the military in automatic gisting (understanding the major intent of a message) technology. The goal of automatic gisting is to automatically gist voice traffic in real-time for the updating of databases and to produce in-time reports. Such a capability will significantly increase the ability to collect and process

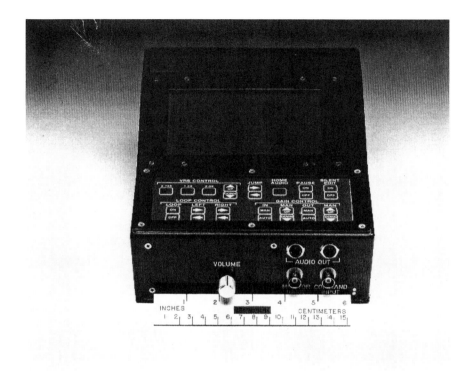

Figure 7. Stand Alone Low Cost Audio Manipulation Unit

large amounts of voice traffic and reduce the data to its most meaningful kernel, i.e., "gist".

However, to develop a gisting technology requires advanced technology in the following areas:

o Continuous Speech Recognition

o Keyword Recognition

o Speaker Identification

o Speaker Adaptation/Normalization

o Natural Language Processing

o Speech Understanding/Artificial Intelligence

o Noise Reduction Techniques

204

- FIVE MINUTES OF DIGITAL AUDIO STORAGE

- SIMULTANEOUS RECORD/PLAYBACK

- JUMP FORWARD/BACKWARD

- LOOP - variable size

- VARIABLE RATE PLAYBACK - 0.65 - 2 x's real-time

- SILENCE/GAP REMOVAL ON PLAYBACK - selectable

- AUTOMATIC PITCH NORMALIZATION

- BANDWIDTH - 100 - 3700 Hz

- SENSITIVITY - 1 millivolt

- DELAY/RESPONSE TIME - 300 millisec. maximum

- COMPACT IN SIZE - 3"H x 7"W x 10"D

- LIGHT WEIGHT - 10 lbs

Figure 8. Specifications on the Audio Manipulation

Unit in Figure 7.

Although much research is being conducted in several of these technology areas independently for commercial and military applications, much less research is being conducted to combine these technologies. Due to the harsh military environment, keyword/phase recognition performance has been very poor. It is, therefore, essential to combine these technologies to obtain a robust gisting capability. The technology is being applied to air traffic control voice communications.

The goal of the research is to extract information from the communication that takes place between the aircraft and the control tower. The system would be capable of producing a gist of the dialog and would compile the information about the transactions and activities that occurred. Some of the desired capabilities are:

o Separate the speech between pilots and controllers

o Determine the airline and flight number

o Identify both the pilot and controller

o Determine the activity underway such as takeoff, landing, etc.

A final goal of the research is to develop a real-time testbed system to perform the extensive testing necessary to assess the current technology as well as provide future direction for research and development to address military field operations.

FUTURE DIRECTION

The use of speech technology for military applications has been shown to increase the effectiveness of a variety of operational tasks. Figure 9 shows some basic speech processing technologies along with the current capabilities and future challenges for these technologies. Speaker identification current performance is given for small numbers of speakers while the challenge is to obtain this performance for a hundred or more speakers. Several new techniques are being pursued to meet the speech enhancement challenge. However, it is crucial that the speech signal not be degraded by the process. Measurement techniques for speech enhancement are not standardized

	TODAY	CHALLENGE
SPEECH RECOGNITION		
VOCABULARY SIZE	300	1000
TYPE OF SPEECH	ISOLATED	CONTINUOUS
ACCURACY	97% (CLEAN)	95% (COCKPIT)
SPEAKER IDENTIFICATION		
ACCURACY	94% (CLEAN)	98% (NOISY)
SPEECH ENHANCEMENT		
NOISE REDUCTION	18 dB	30 dB
SPEECH COMPRESSION		
BITS PER SECOND	2400	50 - 400
JAM RESISTANT FACTOR	2	10 TO 100

Figure 9. Speech Technology Capabilities Today &

the Challenge for the Future

making performance comparison of techniques difficult. New laboratory measurement techniques that better estimate field performance are required. The figure also shows the speech compression jam resistant factor referenced to a four kilohertz bandwidth. The challenge here is to obtain the jam resistant factor and meet intelligibility requirements when the acoustic environmental noise is high such as in the cockpit of an aircraft or in a ground vehicle. It must also be pointed out that minimizing both adaptation and training time is an important and challenging issue for many military applications.

Significant progress has been made in the development of audio signal analysis, voice input, message sorting and automatic gisting technologies. Although several technologies look promising for providing automatic sorting and gisting capabilities for military applications, these technologies can not meet todays requirements. However, the use of these technologies in combination offers a potential solution to improving performance to an acceptable level for use in the field.

In order to provide these speech processing capabilities to the field for test, evaluation, and operation, an increase in processing power per size, weight and cost is required. Therefore, the development of very high speed speech processors that can provide the processing power to support multiple speech functions and channels is essential if these technologies are to meet military requirements and be economically transitioned to both airborne and ground operations.

REFERENCES

[1] Beek, Dr.,B., and Cupples, E.J., et al. "Trends and Application of Automatic Speech Technology." S.D. Harris (ed.) Symposium on Voice-Interactive Systems: Applications and Payoffs, Dallas, TX, 1980.

[2] Beek, Dr.B., and Neuberg, E.P., Hodge, D.C. "An Assesment of the Technology of Automatic Speech Recognition for Military Applications." Acoustic, Speech, and Signal Processing, Aug 1977.

[3] Cupples, E.J., and Foelker, J.L. "Air Force Speech Enhancement Program." Military Speech Tech '87, Vol 1, No. 2, Media Dimensions, Inc., NY, NY, 1987.

[4] Cupples, E.J. "Speech Research and Development at Rome Air Development Center." Military Speech Tech '87, Vol 1, No. 2, Media Dimensions, Inc., NY, NY, 1987.

[5] Desiplio, R.G., and Fry, E. "Avionics System Plays 'Ask and Tell' with Its Operator." Speech Technology, Vol 1, No. 4, 1983.

[6] Lea W.A. "The Value of Speech Recognition Systems." W.A. Lea, (ed.).
     Trends in Speech Recognition, Prentice-Hall, Englewood Cliffs, NJ 1980.

[7] Levinson, A.E., and Liberman, M.Y. "Speech Recognition by Computer."
     Scientific American, Vol 244, No. 4, 1981.

[8] Naylor, Dr., J., Wrench, Dr., E., and Wohlford, R. " Multi - Channel
     Speaker Recognition." RADC Tech. Report Number TR-85-280, 1986.

[9] Simpson, C.A., Coler, C.R., and Huff, E.M. "Human Factors of Voice I/O
     for Aircraft Cockpit Controls and Displays." D.S. Pallet (ed.). Workshop
     on Standardization for Speech I/O Technology, NBS, Gaithersburg, MD,
     1982.

[10] Vonusa, R., Cupples, E.J., et al. "Application, Assesment and
     Enhancement of Speech Recognition for the Aircraft Environment.
     "Advanced Avionics and the Military Aircraft Man/Machine Interface,
     AGARD Proceedings No. 329, France.

[11] Weiss, M., and Aschkenasy, E. " The Speech Enhancement Advanced
     Development Model." RADC Technical Report Number TR-78-232, 1978.

[12] Weiss, M., and Aschkenasy, E. "Wideband Speech Enhancement
     Addition." RADC Technical Report Number TR-81-53. 1981.

[13] Woodard, J.P., and Cupples, E.J. "Selected Military Applications of
     Automatic Speech Recognition Technology." IEEE Communications,
     Vol 21, No. 9, NY, NY, Dec 1983.

[14] Howard, J.A., "Flight Testing of the AFTI/F-16 Voice Interactive Avionics
     System," Proc. Military Speech Tech. 1987, (Media Dimensions),
     Arlington, VA, November 1987.

[15] Hale, M., and Norman, O.D., "An Application of Voice Recognition to
     Battle Management," Proc. Military Speech Tech. 1987 (Media
     Dimensions), Arlington, VA, November 1987.

[16] Naylor, J.A., and Boll, S.F., "Techniques for Suppression of an Interfering
     Talker in Co-Channel Speech," IEEE Proc. ICASSP'87, pp. 205-208,
     April 1987.

[17] Weinstein, C.J., "Opportunities for Advanced Speech Processing in
     Military Computer-Based System" to be published in DARPA Speech and
     Natural Language Workshop.

NOTE: RADC (Rome Air Development Center) was renamed Rome Laboratory during November 1990.

SELECTIVE BIBLIOGRAPHY

This bibliography with abstracts has been prepared by the Scientific and Technical Information Division of the U.S. National Aeronautics and Space Administration, Washington, D.C., in consultation with Editor Prof. Dr. A. N. Ince.

UTTL: Recognition of speaker-dependent continuous speech with KEAL
AUTH: A/MERCIER, G.; B/BIGORGNE, D.; C/MICLET, L.; D/LE
GUENNEC, L.; E/QUERRE, M. PAA: E/(CNET, Lannion, France)
IEE Proceedings, Part I: Communications, Speech and Vision (ISSN
0143-7100), vol. 136, pt. I, no. 2, April 1989, p. 145-154.

ABS: A description of the speaker-dependent continuous speech recognition
system KEAL is given. An unknown utterance, is recognized by means of
the following procedures: acoustic analysis, phonetic segmentation and
identification, word and sentence analysis. The combination of
feature-based, speaker-independent coarse phonetic segmentation with
speaker-dependent statistical classification techniques is one of the main
design features of the acoustic-phonetic decoder. The lexical access
component is essentially based on a statistical dynamic programming
technique which aims at matching a phonemic lexical entry containing
various phonological forms, against a phonetic lattice. Sentence recognition
is achieved by use of a context-free grammar and a parsing algorithm derived
from Earley's parser. A speaker adaptation module allows some of the
system parameters to be adjusted by matching known utterances with their
acoustical representation. The task to be performed, described by its
vocabulary and its grammar, is given as a parameter of the system.
Continuously spoken sentences extracted from a 'pseudo-Logo' language
are analyzed and results are presented. 89/04/00 89A39218

UTTL: Aircrew recommendations for voice message functions in tactical
aircraft
AUTH: A/FOLDS, DENNIS J.; B/BEARD, RODERICK A. PAA:
B/(Georgia Institute of Technology, Atlanta) IN:Human Factors Society,
Annual Meeting, 32nd, Anaheim, CA, Oct. 24-28, 1988, Proceedings.
Volume 1 (A89-31601 12-54). Santa Monica, CA, Human Factors Society,
1988, p. 63-67.

ABS: Results are presented from a survey of 135 active tactical aircrews
regarding use of synthetic voice messages tactical aircraft. The sample was
primarily composed of F-16, F-15, and F-4 pilots. The participants rated 69
existing, proposed, or suggested functions for voice messages in tactical
aircraft. Over two-thirds of the participants rated the following functions
favorably: Engine Fire, Fuel Low, Oil Pressure, Hydraulic Pressure,Brakes
Malfunction, Landing Gear Malfunction, Gear/Faps Configuration, Low
Altitude, Missile Launch, Threa Display, Bingo Fuel, and Joker Fuel. Other
functions, applicable to some but not all tactical aircraft, received strong
support from the aircrews of the applicable aircraft. The participants'
responses to open-ended questions, concerning use of voice messages for

checklists and desirable control features for voice message systems, are also summarized. 88/00/00 89A31613

vol. 36, Oct. 1988, p. 1651-1664. Research supported by the University of California, General Electric Co., NASA, and NSF.

ABS: A family of architectural techniques are proposed which offer efficient computation of weighted Euclidean distance measures for nearest-neighbor codebook searching. The general approach uses a single metric comparator chip in conjunction with a linear array of inner product processor chips. Very high vector-quantization (VQ) throughput can be achieved for many speech and image-processing applications. Several alternative configurations allow reasonable tradeoffs between speed and VLSI chip area required. 88/10/00 89A11388

UTTL: Binaural speech discrimination under noise in hearing-impaired listeners
AUTH: A/KUMAR, K. V.; B/RAO, A. B. PAA: A/(NASA, Johnson Space Center, Houston, TX; Institute of Aviation Medicine, Bangalore, India); B/(Institute of Aviation Medicine, Bangalore, India) CORP: National Aeronautics and Space Administration. Lyndon B. Johnson Space Center, Houston, TX.; Institute of Aviation Medicine, Bangalore (India). Aviation, Space, and Environmental Medicine (ISSN 0095-6562), vol. 59, Oct. 1988, p. 932-936.

ABS: This paper presents the results of an assessment of speech discrimination by hearing-impaired listeners (sensori-neural, conductive, and mixed groups) under binaural free-field listening in the presence of background noise. Subjects with pure-tone thresholds greater than 20 dB in 0.5, 1.0 and 2.0 kHz were presented with a version of the W-22 list of phonetically balanced words under three conditions: (1) 'quiet', with the chamber noise below 28 dB and speech at 60 dB; (2) at a constant S/N ratio of +10 dB, and with a background white noise at 70 dB; and (3) same as condition (2), but with the background noise at 80 dB. The mean speech discrimination scores decreased significantly with noise in all groups. However, the decrease in binaural speech discrimination scores with an increase in hearing impairment was less for material presented under the noise conditions than for the material presented in quiet. 88/10/00 89A11278

UTTL: Current military/government applications for speech recognition
AUTH: A/HICKS, JAMES W., JR. PAA: A/(SCI Technology, Inc., Huntsville, AL) IN: Aerospace Behavioral Engineering Technology Conference, 6th, Long Beach, CA, Oct. 5-8, 1987, Proceedings (A89-10576

01-54). Warrendale, PA, Society of Automotive Engineers, Inc., 1988, p. 37-39.

ABS: This paper presents an overview of several military/government programs in which SCI Technology has implemented and tested its speech recognition technology. Included are the Speckled Trout (U.S. Air Force), LHX

UTTL: Static and dynamic error propagation networks with application to speech coding

AUTH:A/ROBINSON, A. J.; B/FALLSIDE, F. PAA: B/(Cambridge University, England)   IN: Neural information processing systems; Proceedings of the First IEEE Conference, Denver, CO, Nov. 8-12, 1987 (A89-29002 11-63). New York, American Institute of Physics, 1988, p. 632-641.

ABS: Error propagation nets have been shown to be able to learn a variety of tasks in which a static input pattern is mapped onto a static output pattern. This paper presents a generalization of these nets to deal with time varying, or dynamic patterns, and three possible architectures are explored. As an example, dynamic nets are applied to the problem of speech coding, in which a time sequence of speech data are coded by one net and decoded by another. The use of dynamic nets gives a better signal to noise ratio than that achieved using static nets. 88/00/00 089A29049

UTTL: Vector excitation coding with dynamic bit allocation

AUTH:A/YONG, MEI; B/GERSHO, ALLEN PAA: B/(California, University, Santa Barbara) CORP: California Univ., Santa Barbara. IN: GLOBECOM'88 - IEEE Global.

Telecommunications Conference and Exhibition, Hollywood, FL, Nov. 28-Dec. 1, 1988, Conference Record. Volume 1 (A89-26753 10-32). New York, Institute of Electrical and Electronics Engineers, Inc., 1988, p. 290-294. Research supported by NASA, California MICRO Program, and Bell communications Research, Inc.

ABS: Vector excitation coding (VXC) has shown promise for digital transmission of fairly high communications quality speech at low bit rates, but current version of the algorithm still suffer from audible degradations, particularly at 4.8 kb/s. The authors examine the technique of dynamic bit allocation (DBA) to improve the performance of VXC for a given bit rate. The approach is based on the observation that the minimum bit rate needed to code adequately both the long-term and short-term predictors in VXC varies dynamically with time. Therefore, in frames where fewer bits suffice for the predictors, the unneeded bits can be reallocated to other coder parameter sets such as excitation vectors. By dynamically distributing available bits among the different coder parameter sets while keeping the total number of bits in each frame fixed, it is possible to improve overall coder performance without increasing bit rate. 88/00/00 89A26767

UTTL: Systolic architectures for vector quantization
AUTH: A/DAVIDSON, GRANT A.; B/CAPPELLO, PETER R.;
C/GERSHO, ALLEN PAA: A/(Dolby Laboratories, San Francisco, CA); C/(California, University, Santa Barbara) CORP: California Univ., Santa Barbara. IEEE Transactions on Acoustics, Speech, and Signal Processing (ISSN 0096-3518), (Light Helicopter Experimental, U.S. Army), Space Shuttle (NASA), Space Station, AFTI F-16, and ATF (Advanced Tactical Fighter programs. Some of the programs consist of technology demonstrations, while others involve flighttesting, and one, Speckled Trout, operationally installing and utilizing a system on a continual basis. In somecases, the hardware consists of an SCI Voice Control Unit(VCU-5137) and in others, a Voice Development System (VDS-7001)

.RPT# :SAE PAPER 871750 88/00/00 89A10580

UTTL: Smart command recognizer (SCR) - For development, test, and implementation of speech commands
AUTH: A/SIMPSON, CAROL A.; B/BUNNELL, JOHN W.;
C/KRONES, ROBERT R.   PAA: A/(Psyco-Linguistic Research Associates, Woodside, CA); B/(NASA, Ames Research Center; SYRE, Inc., Moffett Field, CA); C/(Sterling Software, Informatics Div., Palo Alto, CA) CORP: Psycho-Linguistic Research Associates, Menlo Park, CA.; National Aeronautics and Space Administration. Ames Research Center, Moffett Field, CA.; Sterling Software, Palo Alto, CA. IN: AIAA, Flight Simulation Technologies Conference, Atlanta, GA, Sept. 7-9, 1988, Technical Papers (A88-53626 23-09). Washington, DC, American Institute of Aeronautics and Astronautics, 1988, p. 215-221.
ABS: The SCR, a rapid prototyping system for the development ,testing and implementation of speech commands in a flight simulator or test aircraft, is described. A single unit performs all functions needed during these three phases of system development, while the use of common software and speech command data structure files greatly reduces the preparation time for successive development phases. As a smart peripheral to a simulation or flight host computer, the SCR interprets the pilot's spoken input and passes command codes to the simulation or flight computer.

RPT#: AIAA PAPER 88-4612   88/00/00   88A53654

UTTL: Generic voice interface for cockpit application
AUTH:A/WILLIAMSON, DAVID T.; B/FEITSHANS, GREGORY L.
PAA: B/(USAF, Wright Aeronautical Laboratories, Wright-Patterson AFB, OH)   IN: NAECON 88; Proceedings of the IEEE National Aerospace and Electronics Conference, Dayton, OH, May 23-27, 1988. Volume 3 (A88-50926 22-01). New York, Institute of Electrical and Electronics Engineers, 1988, p. 780-782.

ABS: A voice technology interface is proposed that would allow   both novice and expert users of voice input and output devices to quickly interface them to their applications while maintaining optimum performance. The objective of this generic voice interface (GVI) is to provide a device-independent interface to existing voice systems . The system will be designed so that any application, not just cockpit applications, can be used with the GVI. Once it has been successfully integrated into a few key applications, the same techniques can be transitioned to other areas. The system will initially be targeted for the rapidly reconfigurable crew-station(RRC) program, which will provide a rapid prototyping environment for advanced crew-station design. 88/00/00 88A50997

UTTL: Neural network classifiers for speech recognition
AUTH:A/LIPPMANN, RICHARD P. PAA: A/(MIT, Lexington, MA)
The Lincoln Laboratory Journal (ISSN 0896-4130), vol. 1, Spring 1988, p. 107-124.

ABS: 'Neural net' classifiers for speech-recognition tasks are presented and compared with conventional classification algorithms. 'Perceptron' neural-net classifiers trained with the novel back-propagation algorith, have been tested and found to yield performance comparable to that of conventional classifiers on digit-classification and vowel-classification tasks. The new 'Viterbi net' architecture, which recognizes time-varying input patterns, furnishes accuracies of the order of 99 percent on a large speech data base. Both perceptron and feature-map neural nets have been implemented on a VLSI device.
RPT#: AD-A203507 ESD-TR-88-255 88/00/00 88A49040

UTTL: Some aspects of automatic speech recognition under helicopter vibration
AUTH:A/RODD, G. M.; B/LEEKS, C. PAA: B/(Royal Aircraft Establishment, Human Engineering Div., Farnborough, England)   IN: Helicopter vibration and its reduction; Proceedings of the Symposium,

London, England, Nov. 16, 1987 (A88-46260 19-05). London, Royal Aeronautical Society, 1987, p. 31-49.

ABS: Attention is given to the problem of helicopter vibration-induced performance degradation in cockpit direct voice input (DVI) control systems. The problem is especially acute at the two resonant frequencies of the larynx. Data have been obtained for DVI of single digits and triple digits; the latter is understandably the moreseverely affected by the 20-Hz vertical vibration condition. Speech recognition is also substantially affected and calls for additional helicopter noise-reduction efforts. 87/00/00 88A46263

UTTL: Acoustic-phonetic changes in speech due to environmental stressors - Implications for speech recognition in the cockpit

AUTH: A/MOORE, THOMAS J.; B/BOND, Z. S. PAA: A/(USAF, Harry G. Armstrong Aerospace Medical Research Laboratories, Wright-Patterson AFB, OH); B/(Ohio University, Athens) IN: International Symposium on Aviation Psychology, 4th, Columbus, OH, Apr. 27-30, 1987, Proceedings (A88-42927 17-53). Columbus, OH, Ohio State University, 1987, p. 77-83.appears to be a sensitive and reliable instrument for measuring speech recognition algorithm performance. 87/00/00 88A34144

UTTL: The development and status of a robust speech recognition data base

AUTH: A/ERICSON, MARK A. PAA: A/(USAF, Armstrong Aerospace Medical Research Laboratory, Wright-Patterson AFB, OH) IN: NAECON 87; Proceedings of the IEEE National Aerospace and Electronics Conference, Dayton, OH, May 18-22, 1987.Volume 3 (A88-34026 13-01). New York, Institute of Electrical and Electronics Engineers, Inc., 1987, p. 882-888.

ABS: The author describes the development of the DARPA robust speech database, the facilities used to collect it, the simulated environmental conditions, the database configurations, the progress and the status at the present time. Preselected speech is produced by experienced aircraft pilots under simulated harsh fighter aircraft environmental conditions. This stressed speech comprises the database that can serve to provide benchmark performance of speech recognition systems and to further the research of efforts of environmental stressors on speech. 87/00/00 88A34143

UTTL: Speaking to military cockpits

AUTH:A/WHITE, R. G.   PAA: A/(Royal Aircraft Establishment,   Bedford,
England)   IN: Recent advances in cockpit aids for military operations;
Proceedings of the Symposium, London, England, Mar. 31, 1987
(A88-32676 12-01). London, Royal Aeronautical Society, 1987, p. 74-88.

ABS: The extent to which the theoretical benefits of Direct Voice Input can be
realized in practice has been evaluated in the UK by three flight test-based
studies and one flight simulator-based study. The recognition error rates
recorded during the four studies are compared with a target recognition
performance level that is regarded as the minimum required for operational
use. It is concluded that, while measured error rates are currently too high,
continued research will allow a proposed timetable for entry into service to
be met. 87/00/00 88A32682

UTTL: Automatic voice alert devices (AVAD)
AUTH:A/MARSHALL, R.   PAA: A/(Racal Acoustics, Ltd., Wembley,
England)   IN: Recent advances in cockpit aids for military operations;
Proceedings of the Symposium, London, England, Mar. 31, 1987
(A88-32676 12-01). London, Royal Aeronautical Society, 1987, p. 70-73.

ABS: Automatic Voice Alert Devices (AVADs) give warnings of   aircraft
malfunctions to crews in unmistakable human speech, improving reaction
time and reducing crew stress. An attention-getting tone, or 'attenson'
precedes each message. Attention is presently given to the

ABS: The effects of various environmental stressors, such as noise, oxygen
mask, acceleration, and vibration, on speech production were investigated.
Recordings of speech of male speakers, who wore standard Air Force flight
helmets with oxygen masks and were breathing air supplied through a
chest-mounted regulator, were made during centrifuge rides. It was found
that, compared to control conditions, fundamental frequency increased
under all experimental conditions for all talkers, with each of the stressors
resulting in the increased vocal effort of the talker (reflected in an increase
in fundamental frequency). Relative amplitude increased for speech
produced in the presence of noise when a boom microphone was used, but
showed no systematic change when the talker wore an oxyge mask, as was
the case for speech produced under acceleration and vibration when the
talkers wore oxygen masks. In general, the vowel space became more
compact for speech produced in presence of any of the stressors. 87/00/00
88A42938

UTTL: Speech technology in the flight dynamics laboratory
AUTH:A/WILLIAMSON, DAVID T.; B/SMALL, RONALD L.;
C/FEITSHANS, GREGORY L.   PAA: C/(USAF, Wright Aeronautical

Laboratories, Wright-Patterson AFB, OH) IN: NAECON 87; Proceedings of the IEEE National Aerospace and Electronics Conference, Dayton, OH, May 18-22, 1987. Volume 3 (A88-34026 13-01). New York, Institute of Electrical and Electronics Engineers, Inc., 1987. p. 897-900.

ABS: Over the past several years, the flight dynamic laboratory at Wright-Patterson Air Force Base has been actively involved in the investigation of the role of speech technology in US Air Force cockpits. The authors provide a summary of progress to date and also discuss the future direction of speech technology applicationsresearch within the flight dynamics laboratory. 87/00/00 88A34145

UTTL: Phonetic discrimination (PD-100) test for robust speech recognition
AUTH:A/SIMPSON, CAROL A.; B/RUTH, JOHN C. PAA: A/(Psycho-Linguistic Research Associates, Woodside, CA); B/(McDonnell Douglas Electronics Co., Saint Charles, MO) IN: NAECON 87; Proceedings of the IEEE National Aerospace and Electronics Conference, Dayton, OH, May 18-22, 1987.Volume 3 (A88-34026 13-01). New York,Institute of Electrical and Electronics Engineers, Inc., 1987, p. 889-896.

ABS: A rigorous phonetic discrimination test that can be used to compare recognizers and to predict the performance of individual recognizers for specific operational vocabularies is described. Preliminary results obtained with two connected word speaker-dependent recognizers compared to a human listener are reported. On the basis of these preliminary data, the phonetic discrimination test characteristics of the V694 AVAD, which is activated by keying signals from remote sensor systems and holds about 64 sec of digitally-recorded human speech and attensons; the unit memory of tones, words, and phrases can be controlled by software to generate a total message output far exceeding the vocabulary storage. 87/00/00 88A32681

UTTL: Automatic voice identification system
AUTH:A/WOODSUM, HARVEY PAA: A/(Sanders Associates, Engineering Div., Merrimack, NH) Lockheed Horizons (ISSD 0459-6773), Dec. 1987, p. 12-17.

ABS: Advances in the basic science of voice identification when combined with the latest state-of-the-art digital electronics are discussed. The Sanders voice identification system prototype consisting of an audio input device, a microcomputer, and special hardware for Computing spectrographic features is described. The Griffin Pattern classifier (GPC) for computing the likelihood of each identity is discussed. Results of the experimental testing of the system in which 30-sec samples of seven voices were used are listed. Such applications as controlling access to computer data files are discussed. 87/12/00 88A26418

UTTL: The use of speech technology in air traffic control simulators
AUTH:A/HARRISON, J. A" B/HOBBS, G. R.; C/HOWES, J. R.; D/COPE, N. PAA: D/(Ferranti Computer Systems, Ltd., Bracknell, England) IN: International Conference on Simulators, 2nd, Coventry, England, Sept. 7-11, 1986, Proceedings (A88-16676 04-09). London, Institution of Electrical Engineers, 1986, p. 15-19.
ABS: The advantages of applying speech technology to air traffic control (ATC) simulators are discussed with emphasis placed on the simulation of the pilot end of the pilot-controller dialog. Speech I/0 in an ATC simulator is described as well as technology capability, and research on an electronic blip driver. It is found that the system is easier to use and performs better for less experienced controllers. 86/00/00 88A16678

UTTL: The graph search machine (GSM) - A VLSI architecture for connected speech recognition and other applications
AUTH:A/GLINSKI, STEPHEN C.; B/LALUMIA, T. MARIANO; C/CASSIDAY, DANIEL R.; D/KOH, TAIHO; E/GERVESHI, CHRISTINE PAA: E/(AT&T Bell Laboratories, Murray Hill, NJ) IEEE, Proceedings (ISSN 0018-9219), vol. 75, Sept. 1987, p. 1172-1184.
ABS: A programmable VLSI architecture is described for efficiently computing a variety of kernel operations for speech recognition. These operations include dynamic programming for isolated and connected word recognition. Using both the template matching approach and the hidden markov model (HMM) approach, the use of finite-state grammars (FSG) for connected word recognition, and metric computations for vector quantization and distance measurement. These are collectively referred to as 'graph search' operations since a diagram consisting of arcs and nodes is commonly used to illustrate the HMM or FSG. As well as being able to efficiently compute a wide class of speech processing operations, the architecture is useful in other areas such as image processing. A chip design has been completed using 1.75-micron CMOS design rules and combines both custom and standard cell approaches. 87/09/00 88A15387

UTTL: Gain-adaptive vector quantization with application to speech coding
AUTH:A/CHEN,JUIN-HWEY; B/GERSHO, ALLEN PAA: A/(Code)

Corp., Mansfield, MA); B/(California, University, Santa Barbara) CORP: Codex Corp., Mansfield, MA.; California Univ., Santa Barbara.IEEE Transactions on Communications (ISSN 0090-6778), vol. COM-35, Sept. 1987, p. 918-930. Research supported by the State of California MICRO Program, General Electric Co., and NASA.

ABS: The generalization of gain adaptation to vector quantization (VQ) is explored in this paper, and a comprehensive examination of alternative techniques is presented. A class of adaptive vector quantizers that can dynamically adjust the 'gain' or amplitude scale of code vectors according to the input signal level is introduced. The encoder uses a gain estimator to determine a suitable normalization of each input vector prior to VQ encoding. The normalized vectors have reduced dynamic range and can then be more efficiently coded. At the receiver, the VQ decoder output is multiplied by the estimated gain. Both forward and backward adaptation are considered, and several different gain estimators are compared and evaluated. Gain-adaptive VQ can be used alone for 'vector PCM' coding (i.e., direct waveform VQ) or as a building block in other vector coding schemes. The design algorithm for generating the appropriate gain-normalized VQ codebook is introduced. When applied to speech coding, gain-adaptive VQ achieves significant performance improvement over fixed VO with a negligible increase in complexity. 87/09/00 88A11171

UTTL: Versatile simulation testbed for rotorcraft speech I/0 system design
AUTH: A/SIMPSON, CAROL A. PAA: A/(Psycho-Linguistic Research Associates, Menlo Park, CA) CORP: Psycho-Linguistic Research Associates, Menlo Park, CA. IN: Aerospace Behavioral Engineering Technology Conference, 5th, Long Beach, CA, Oct. 13-16, 1986, Proceedings (A88-10152 01-54). Warrendale, PA, Society of Automotive Engineers, Inc., 1986, p. 33-37. USAF-supported research. Multiclique) and presents also some considerations on the problem of transition from 64 kbit/s TDMA/DSI system to 32 kbit/s TDMA/DSI system. 86/00/00 87A49892

UTTL: DVI in the military cockpit - A third hand for the combat pilot
AUTH:A/WANSTALL, BRIAN Interavia (ISSN 0020-5168), vol. 42., June 1987, p. 655, 656, 659. 660.
ABS: Voice warning systems are already in service, and advancements in both hardware and software development promise usable automatic speech recognition (ASR) systems for next-generation combat aircraft and battlefield helicopters. Direct voice input (DVI) ASR systems have been specifded for next-generation NATO fighters and combat helicopters, with

the aim of reducing overall workload and stress levels. Noise, g-levels and pressurized breathing, however, militate against the easy incorporation of DVI; verification or feedback of the voice commands in some yet to be defined form is noted to be vital for successful combat use.   87/06/00 87A46315

UTTL: 16 kb/s high quality voice encoding for satellite communication networks
AUTH:A/YATSUZUKA, YDHTARO; B/YAMAZAKI, TOMOHIRO; C/IIZUKA, SHIGERU   PAA: C/(Kokusai Denshin Denwa Co., Ltd., Tokyo, Japan)   International Journal of Satellite Communications (ISSN 0737-2884), vol. 4, Oct.-Dec. 1986, p. 193-202.
ABS: A 16 kb/s adaptive predictive coding (APC) with maximum   likelihood quantization (MLQ), which can cover a range of coding rates from 4.8-16 kb/s, for low C/N satellite communications systems is described, and its performance is evaluated. The requirements for a 16 kb/s voice coding technique in low C/N digital satellite communication systems, such as maritime and thin-route communications, are discussed. The use of a 9.6 kb/s voice coding channel for small-size antenna systems is proposed. NEC-7720 DSP chips were employed to implement the 16 kb/s APC/MLQ codec. A multimedia multiplexing for low C/N digitalcommunications systems, and a small-scale circuit multiplication system for business services are examined. It is observed that the 16 kb/s APC hardware code with MLQ is applicable for speech and nonvoice signals.   86/12/00 87A37831

UTTL: Research on speech processing for military avionics
AUTH: A/MOORE, THOMAS J.; B/MCKINLEY, RICHARD L. PAA: B/(USAF, Armstrong Aerospace Medical Research Laboratory, Wright-Patterson AFB, OH)   IN : Human Factors Society, Annual Meeting, 30th, Dayton, OH, Sept. 29-Oct. 3, 1986,Proceedings. Volume 2 (A87-33001 13-54). Santa Monica, CA, Human Factors Society, 1986, p. 1331-1335.
ABS: The Biological Acoustics Branch of the Armstrong Aerospace Medical Research Laboratory (AAMRL) is engaged in research
ABS: A versatile simulation testbed for the design of a rotorcraft speech I/O system is described in detail. The testbed will be used to evaluate alternative implementations of synthesized speech displays and speech recognition controls for the next generation of Army helicopters including the LHX. The message delivery logic is discussed as well as the message structure, the speech recognizer command structure and features, feedback from the recognizer, and random access to controls via speech command.

RPT#:SAE PAPER 861661   86/00/00   88A10154

UTTL: Continuous speech recognition on a Butterfly Parallel Processor
AUTH:A/COSELL, LYNN; B/KIMBALL, OWEN; C/SCHWARTZ,
RICHARD; D/KRASNER, MICHAEL   PAA: D/(BBN Laboratories,
Inc., Cambridge, MA)   IN: 1986 International ConferenCe on Parallel
Processing, University Park, PA, Aug. 19-22, 1986, Proceedings (A87-52528
24-62). Washington, DC, IEEE Computer Society Press, 1986, p. 717-720.
DARPA-sponsored research.

ABS: This paper describes the implementation of a continuous speech
recognition algorithm on the BBN Butterfly Parallel Processor. The
implementation exploited the parallelism inherent in the recognition
algorithm to achieve good performance, as indicated by execution time and
processor utilization. The implementation process was simplified by a
programming methodology that complements the Butterfly architecture.
The paper describes the architecture and methodology used and explains
the speech recognition algorithm, detailing the computationally demanding
area critical to an efficient parallel realization. The step taken to first
develop and then refine the parallel implementation are discussed, and the
appropriateness of the architecture and programming methodology for such
speech recognition applications is evaluated.   86/00/00 87A52611

UTTL: Low rate encoding - A means to increase system capacity in a
TDMA/DSI system
AUTH:A/FRESIA, F.; B/PATACCHINI, A.; C/PRIN5, C. PAA:
C/(EUTELSAT, Paris, France) IN: ICDSC-7; Proceedings of the Seventh
International Conference on Digital Satellite Communications, Munich,
West Germany, May 12-16, 1986 (A87-49886 22-32). Berlin, West Germany,
VDE-Verlag GmbH, 1986, p. 81-88.

ABS: The satellite capacity required in a TDMA/DSI system for a certain
amount of traffic, can be substantially reduced when using associated LRE
techniques. An analysis based on the future Eutelsat TDMA/DSI traffic
forecast has been made in this respect, by using the computer program
presently in use at EUTELSAT for the generation of Burst Time Plans. The
analysis considers different operational modes (Single Destination ,
Multidestination and in a number of speech related areas. This paper
describes the approach used to conduct research in the development and
evaluation of military speech communication systems, mentions the types of
studies done using this approach, and gives examples of the types of data
generated by these studies. Representative data are provided describing

acoustic-phonetic changes that occur when speech is produced under acceleration. 86/00/00 87A33070

UTTL: Recognition of synthesized, compressed speech in noisy environments
AUTH: A/GARDNER, DARYLE JEAN; B/BARRETT BRYAN; C/BONNEAU. JOHN ROBERT; D/DOUCET, KAREN; E/IANDERWEYDEN, PROSPER PAA: E/(Kearney State College, NE) IN: Human Factors Society, Annual Meeting, 30th Dayton, OH, Sept. 29-Oct. 3, 1986, Proceedings. Volume 2 (A87-33001 13-54). Santa Monica, CA. Human Factors Society, 1986, p. 927-930.

ABS: The purpose of the present study was to investigate the recognition of synthesized, compressed speech under helicopter noise vs. ambient noise conditions. Subjects performed an isolated word recognition task for stimuli generated by the VOTAN V-5000A speech synthesizer/recognizer. Results indicated that recognition performance, both in terms of percentage correct and average response time deteriorated as a function of speech compression and level of noise. implications of these results for the employment of compressed, synthesized speech warning systems is rotary wing aircraft are discussed. 86/00/00 87A33049

UTTL: Integrating speech technology to meet crew station design requirements
AUTH: A/SIMPSON, CAROL A.; B/RUTH, JOHN C.; C/MOORE, CAROLYN A. PAA: A/(Psycho-Linguistic Research Associates Menlo Park, CA); B/(McDonnell Douglas Electronics Co., Saint C/(VERAC, Inc., San Diego. CA) IN: Charles, M; Digital Avionics Systems Conference, 7th, Fort Worth, TX, Oct. 13-16, 1986, Proceedings (A87-31451 13-01). New York, Institute of Electrical and Electronics Engineers, Inc., 1986, p. 324-329.

ABS: The last two years have seen improvements in speech generation and speech recognition technology that make speech I/0 for crew station controls and displays viable for operational systems. These improvements include increased robustness of algorithm performance in high levels of background noise, increased vocabulary size, improved performance in the connected speech mode, and less speaker dependence. This improved capability makes possible far more sophisticated user interface design than was possible with earlier technology. Engineering. linguistic and human factors design issues are discussed in the context of current voice I/0 technology performance. 86/00/00 87A31491

UTTL: AFTI/F-16 voice interactive avionics
AUTH:A/ROSENHOOVER, F. A. PAA: A/(General Dynamics Corp.,
Fort Worth, TX) IN: NAECON 1986; Proceedings of the National
Aerospace and Electronics Conference, Dayton, OH, May 19-23, 1986.
Volume 2 (A87-16726 05-01). New York, Institute of Electrical and
Electronics Engineers, 1986, 613-617.

ABS: This paper discusses the integration of voice in a fighter environment,
including work load assessments and the approach used to solve the work
load demands on the pilot. Tasks within the crew station are identified
according to their ability to increase the pilot's awareness of his environment
and ability to maintain his mission objectives with minimal error. A
discussion of the areas of evaluation during various mission profiles is
presented to establish interactive needs for mission success. 86/00/00
87A16789

UTTL: Gain-adaptive vector quantization for medium-rate speech coding
AUTH:A/CHEN, J.-H.; B/GERSHO, A. PAA: B/(California University,
Santa Barbara) CORP: California Univ., Santa Barbara. IN: ICC '85;
International Conference on Communications, Chicago, IL, June 23-26,
1985, Conference Record. Volume 3 (A86-37526 17-32). New York,
Institute of Electrical and Electronics Engineers, Inc., 1985, P.
1456-1460.

ABS: A class of adaptive vector quantizers (VQs) that can dynamically
adjust the 'gain' of codevectors according to the input signal level is
introduced. The encoder uses a gain estimator to determine a suitable
normalization of each input vector prior to VO coding. The normalized
vectors have reduced dynamic range and can then be more efficiently coded.
At the receiver, the VQ decoder output is multiplied by the estimated gain.
Both forward and backward adaptation are considered and several different
gain estimators are compared and evaluated. An approach to optimizing the
design of gain estimators is introduced. Some of the more obvious
techniques for achieving gain adaptation are substantially less effective than
the use of optimized gain estimators. A novel design technique that is
needed to generate the appropriate gain-normalized codebook for the
vector quantizer is introduced. Experimental results show that a significant
gain in segmental SNR can be obtained over nonadaptive VQ with a
negligible increase in complexity. 85/00/00 86A37579

UTTL: Continuous speech recognition using natural language constraints
AUTH: A/ROUTH, R. L.; B/MILNE, R. W. PAA: B/(USAF, Institute of

Technology, Wright-Patterson AFB, OH)   IN: NAECON 1984;
Proceedings of the National Aerospace and Electronics Conference,
Dayton, OH, May 21-25, 1984. computer will recognize his voice.
Commands can then call up flight status data, select radio frequencies, and
engage or disengage the autopilot. 84/05/05 84A37038

UTTL: Maximum likelihood spectral estimation and its application to
narrow-band speech coding
AUTH:A/MCAULAY, R. J. PAA: A/(MIT, Lexington, MA) IEEE
Transactions on Acoustics, Speech, and Signal Processing (ISSN
0096-3518), vol. ASSP-32, April 1984, p. 243-251. USAF-sponsored
research.
ABS: Itakura and Saito used the maximum likelihood (ML) method to derive
a spectral matching criterion for autoregressive (i.e., all-pole) random
processes. Their results are generalized to periodic processes having
arbitrary model spectra. For the all-pole model, Kay's covariance domain
solution to the recursive ML (RML) problem is cast into the spectral domain
and used to obtain the RML solution for periodic processes. When applied
to speech, this leads to a method for solving the joint pitch and spectrum
envelope estimation problems. It is shown that if the number of frequency
power measurements greatly exceeds the model order, then the RML
algorithm reduces to apitch-directed, frequency domain version of linear
predictive (LP) spectral analysis. Experiments on a real-time vocoder
reveals that the RML synthetic speech has the quality of being heavily
smoothed.
RPT# : AD-A147582 ESD-TR-84-106 84/04/00 84A32088

UTTL: Application of 32 and 16 kb/s speech encoding techniques to digital
satellite communications
AUTH:A/YATSUZUKA, Y.; B/YATO, F.; C/KUREMATSU, A. PAA:
C/(Kokusai Denshin Denwa Co., Ltd., Research and Development
Laboratories, Tokyo, Japan) (COMSAT, INTELSAT, AIAA, and IEEE,
International Conference on Digital Satellite Communications, 6th,
Phoenix, AZ, Sept. 19-22, 1983) International Journal of Satellite
Communications (ISSN 0737-2884), vol. 1, Oct.-Dec. 1983, p. 113-122.
Research supported by the International Maritime Satellite Organization.
83/12/00 84A31356

UTTL: Experience with speech communication in packet networks
AUTH: A/WEINSTEIN, C. J.; B/FORGIE, J. W. PAA: B/(MIT,

Lexington, MA)   IEEE Journal on Selected Areas in Communications (ISSN 0733-8716), vol. SAC-1, Dec. 1983, p. 963-980. DARPA-supported research.

ABS: It is pointed out that packet techniques provide powerful mechanisms for the sharing of communication resources among users with time-varying demands. The primary application of packet techniques has been for digital data communications. Packet techniques offer significant benefits for voice and. for data. Packet speech concepts and issues are considered, taking into account the generic Volume 2 (A85-44976 21-01). New York, IEEE, 1984, p. 916-923.

ABS: The role of semantic and syntactic constraints in the process of speech recognition is investigated, and a real time, general solution to the application of English syntactic constraints to spoken English recognition is developed that is subject to the accuracy of the acoustic analyzer and the accuracy and completeness of an English Parser. It is noted that automated speech recognition at the level of conversation or dictation (as required in future aircraft cockpit systems) must incorporate several hierarchical levels of sophisticated, artificially intelligent, syntactic and semantic analysis, in addition to the extremely accurate 'front end' word-level acoustic analyzer assumed from the outset. 84/00/00 85A45104

UTTL: Vocoders in mobile satellite communications
AUTH:A/KRIEDTE, W.; B/CANAVESIO, F.; C/DAL DEGAN, N.; D/PIRANI, G.; E/RUSINA, F.; F/USAI, P.   PAA: A/(ESA, Payload Technology Dept., Noordwijk, Netherlands); F/(Centro Studi e Laboratori Telecomunicazioni S.p.A., Turin, Italy)   ESA Journal (ISSN 0379-2285), vol. 8, no. 3, 1984, p. 285-305. Sponsorship European Space Agency.

ABS: Owing to the power constraints that characterize onboard transmission sections, low-bit-rate coders seem suitable for speech communications inside mobile satellite systems. Vocoders that operate at rates below 4.8 kbit/s could therefore be a desirable solution for this application, providing also the redundancy that must be added to cope with the channel error rate. After reviewing the mobile-satellite-systems aspects, the paper outlines the features of two different types of vocoders that are likely to be employed, and the relevant methods of assessing their performances. Finally, some results from computer simulations of the speech transmission systems are reported. 84/00/00 85A17099

UTTL: Voice control on military aircraft
AUTH:A/MELOCCO, J.-M. PAA: A/(Crouzet, 5.A., Valence, Drome,

France) Air et Cosmos (ISSN 0044-6971), May 5, 1984, p. 151-154, 158. In French.

ABS: Progress in introducing voice controls and annunciators in military aircraft to reduce the pilot workload is explored. French work in cockpit voice capability began in 1978 and concentrated initially on calling up data displays. The Mirage III was equipped with a voice annunciator that alerted the pilot to abnormal system functions. The voice was produced completely artificially. Pilot voices were then investigated in simulated conditions of altitudes, vibrations, and accelerations to encode sufficient recognizance programs for on-board computers to understand 100-200 pilot spoken word commands. In the most current system, the pilot must pronounce each command word before flying so that the packet speech system configuration, the generic packet voice terminal configuration, digital speech processing functions, packet speech protocol functions, speech packetization and reconstitution, conferencing techniques, and statistical multiplexing of packet voice and data. Attention is given to a summary of packet speech experiments, packet speech on the ARPA network, packet speech on the Atlantic packet satellite network, and packet speech on the experimental wide-band system.

RPT# : AD-A147058 E50-TR-84-234 83/12/00 84A29906

UTTL: A 500-800 bps adaptive vector quantization vocoder using a perceptually motivated distance measure AUTH: A/PAUL, D. B. PAA: A/(MIT, Lexington, MA)IN: Globecom '82 - Global Telecommunications Conference, Miami, FL, November 29-December 2, 1982, Conference Record. Volume 3 (A84-26401 11-32). New York, Institute of Electrical and Electronics Engineers, 1982, p. 1079-1082. USAF-sponsored research.

ABS: This paper presents a vector quantization system based upon the Spectral Envelope Estimation vocoder. In order to optimize performance, the system employs a perceptually-motivated spectral distance measure and updates the template set continuosly during operation to adapt to the current speaker(s) and environment(s). Strategies have been devised for transmission of new templates on several different communications channels. The system achieves an intelligibility score (DRT) of 86.3 percent for unrehearsed speech on a 760 bps circuit-switched channel. 82/00/00 84A26459

UTTL: Note on the properties of a vector quantizer for LPC coefficients AUTH: A/RABINER, L.R.; B/SONDHI, M. M.; C/LEVINSON, 5. E.

PAA: C/(Bell Telephone Laboratories, Inc., Murray Hill. NJ) Bell System Technical Journal (ISSN 0005-8580), vol. 62, Oct. 1983, p. 2603-2616.

ABS: The results of a series of experimental evaluations of the single-split and binary-split algorithms For vector quantization (VQ) training are discussed. Each of the different splitting criteria leads to a different reference prototype set or VQ code book; however, all the VQ sets have essentially the same average distortion. The coverage of the linear predictive coding space for all VQ sets is identical, and the average distance of any one VQ set from another VQ set is smaller than the average distortion of the training set. Hence, the different implementations of the training algorithm for the VQ lead to equivalent VQ reference sets, and for any practical application the simple binary-split algorithm is effective for deriving the VQ code book entries . The implementation by Linde et al . (1980) of the binary split VQ training algorithm is reviewed and its modification for the ,single - split case is shown . 83/ 10/0084A 18764

UTTL: Speech recognition on a distributed array processor
AUTH:A/SIMPSON, P.; B/ROBERTS, J. B. G.   PAA: B/(Royal Signals and Radar Establishment, Malvern, Worcs., England) Electronics Letters (ISSN 0013-5194), vol: 19, Nov. 24, 1983, p. 1018-1020.

ABS: A highly parallel single-instruction multiple-data array signal processor is advocated as efficient for a wide range of real-time problems. Its performance for digital speech recognition is examined and it is shown that impressive throughput rates for realistic vocabulary sizes can be achieved for 'time-warping' dynamic programming algorithms which currently form the basis of several commercial and research speech recognizers. 83/11/24 84A17228

UTTL: Real-time speech coding
AUTH:A/CROCHIERE, R. E.; B/COX, R. V.; C/JOHNSTON, J. D.
PAA: C/(Bell Telephone Laboratories, Inc., Murray Hill, NJ)   IEEE Transactions on Communications, vol. COM-30, Apr. 1982, p. 621-634.

ABS: This paper reviews recent efforts in the design and implementation of real-time speech coders. The approach and methodology for real-time hardware for coder techniques ranging from low to high complexity are discussed. Examples of realizations are given for each approach. They include adaptive differential PCM coding, subband coding, harmonic scaling with subband coding, and adaptive transform coding. Low to medium complexity techniques are based on the use of a digital signal processing integrated circuit. High complexity block processing techniques are based on the use of an array processing computer. An assessment of the performance versus complexity tradeoffs devolved in these coding methods is given in conclusion. 82/04/00 82A30933

UTTL: Voice recognition and artificial intelligence in an air traffic control environment

AUTH: A/HALL, ROBERT F.   CORP: Air Force Inst. of Tech., Wright-Patterson AFB, OH.

ABS: The rapid growth of air carrier, general aviation, and military traffic has strained this nation's Air Traffic Control (ATC) system. The symptoms of this strain appear as controller fatigue, low controller moral, and the occasional creation of a hazardous situation caused by human error. The current method employed to improve the ATC system has been in the form of increasing its air traffic handling capacity by adding more machinery and manpower. Thus, machines with greater processing power and more humans are coupled into a man machine system which is destined to continually grow. Little has been done to find new forms of technology to increase the joint efficiency of man and machine. Two relatively new technologies which to allow for compression of the words in fluent speech, or had a tightly specified transition matrix, to discourage insertion errors, the results are disappointing. The embedded training procedure improves performance, if the Bakis model structure is used; if a full upper-triangular transition probability matrix is used, the performance is far worse than in the isolated-word training case. If the Baum-Welch algorithm is performed after the segmentation procedure, performance improves, but whether this improvement is sufficient to justify the increase in computer time required is questionable.

RPT# : RSRE-MEMO-4099 BR 104991 ETN-88-92517 AD-A193651   87/12/01 88N23928

UTTL: Algorithms and architectures for acoustic phonetical detection of continuous speech

AUTH:A/VICARD,DOMINIQUE CORP:Ecole Nationale Superieure des Telecommunications, Paris (France).   CSS: (Dept. Electronique et Signal.)

ABS: The algorithm analysis includes choice of parameters, vector quantification, dynamic programming, and transient detection. The integrated circuit architecture analysis includes the solutions to the implementation of vector quantification, dynamic programming, and selection. The CMOS implementation of the integrated circuit is described. The cost/performance ratio of the described device is judged excellent.

RPT#: ENST-87E016 ISSN-0751-1353 ETN-88-9217887/10/00 88N23054

UTTL: Evaluating the performance of the LPC (Linear Predictive Coding) 2.4 kbps (kilobits per second) processor with bit errors using a sentence verification task

AUTH: A/SCHMIDT-NIELSEN, ASTRID; B/KALLMAN, HOWARD J.

CORP: Naval Research Lab., Washington, DC.

ABS: The comprehension of narrowband digital speech with bit errors was tested by using a sentence verification task. The use of predicates that were either strongly or weakly related to the subjects (e.g., A toad has warts./ A toad has eyes.) varied the difficulty of the verification task. The test conditions included unprocessed and processed speech using a 2.4 kb/s (kilobits per second) linear predictive coding (LPC) voice processing algorithm with random bit error rates of 0 percent, 2 percent, and 5 percent. In general, response accuracy decreased and reaction time increased with LPC processing and with increasing bit error rates. Weakly related true sentences and strongly related false sentences were more difficult than their counterparts. Interactions between sentence type and speech processing conditions are discussed.

RPT# : AD-A188573 NRL-9089 87/11/30 88N19686

could create a path towards greater system efficiency are the technologies of voice recognition and artificial intelligence. With greater system efficiency, less controller fatigue and better air safety are expected. Where to apply these technologies, in what form, and how deep these technologies can be integrated into the ATC system are questions which deserve inquiry. This research details a method to answer these questions, develops prototype equipment from which to experiment, and establishes a basis from which other research efforts may be launched. A review of literature indicates that current efforts at applying voice recognition in flight operations are centered around pilot task improvement and special projects such as the space shuttle.

RPT# :AD-A197219 AFIT/CI/NR-88-171 88/05/00 89N12559

UTTL: Analysis and improvement of the vector quantization in SELP (Stochastically Excited Linear Prediction)

AUTH:A/KLEIJN, W. B.; B/KRASINSKI, D. J.; C/KETCHUM, R. H.

CORP: Bell Telephone Labs., Inc., Naperville, IL. In Jet Propulsion Lab., Proceedings of the Mobile Satellite Conference p 527-532 (SEE N88-25680 19-32)

ABS: The Stochastically Excited Linear Prediction (SELP) algorithm is described as a speech coding method employing a two-stage vector quantization. The first stage uses an adaptive codebook which efficiently encodes the periodicity of voiced speech, and the second stage uses a stochastic codebook to encode the remainder of the excitation signal. The adaptive codebook performs well when the pitch period of the speech signal is larger than the frame size. An extension is introduced, which increases its performance for the case that the frame size is longer than the pitch period. The performance of the stochastic stage, which improves with frame length, is shown to be best in those sections of the speech signal where a high level of short-term correlations is present. It can be concluded that the SELP algorithm performs best during voiced speech where the pitch period is longer than the frame length. 88/05/00 88N25756

UTTL: Experimental evaluation of algorithms for connected speech recognition using hidden Markov models
AUTH:A/COOK, ANNELIESE CORP: Royal Signals and Radar Establishment, Malvern (England).
ABS: A method of extracting training utterances from fluent speech and constructing Hidden Markov Models (HMMs) from these templates, known as embedded training, investigated with a two-level algorithm for connected word recognition. The effects on recognition performance of various HMM training procedures are discussed, and experimental results for native and non-native English speakers are presented. Training on isolated words does not produce models adequate for use in connected word recognition; whether the model was highly unconstrained,

UTTL: Speech recognition: Acoustic-phonetic knowledge acquisition and representation
AUTH:A/ZUE, VICTOR W. CORP: Massachusetts Inst. of Tech., Cambridge. CSS: (Research Lab. of Electronics.)
ABS: A long-term research goal is the development and implementation of speaker-independent continuous speech recognition systems. It is believed that the proper utilization of speech-specific knowledge is essential for such advanced systems. Research is thus directed toward the acquisition of acoustic-phonetic and lexical knowledge, and the application of this knowledge to speech recognition algorithms. Investigation into the contextual variations of speech sounds has continued, emphasizing the role of the syllable in these variations. Analysis revealed that the acoustic realization of a stop dependsgreatly on its position within a syllable. In order to represent and utilize this information in speech recognition, a hierarchical syllable description has been adopted that enables us to specify the constraints in terms of an immediate constituent grammar. We will continue to quantify the effect of context on the acoustic realization of phonemes

using larger constituent units such as syllables. In addition, a grammar will be developed to describe the relationship between phonemes and acoustic segments, and a parser that will make use of this grammar for phonetic recognition and lexical access.
RPT#: AD-A187293 87/09/24 88N17893

UTTL: Effect of audio bandwidth and bit error rate on PCM, ADPCM and LPC speech Coding algorithm intelligibility
AUTH:A/MCKINLEY, RICHARD L.; B/MODRE, THOMAS J.
CORP: Aerospace Medical Research Labs., Wright-Patterson AFB, OH.
In AGARD, Information Management and Decision Making in Advanced Airborne Weapon Systems 7 p (SEE N87-29503 24-06)

ABS: The effects of audio bandwidth and bit error rate on speech intelligibility of voice coders in noise are described and quantified. Three different speech coding techniques were investigated, pulse code modulation (PCM), adaptive differential pulse code modulation (ADPCM), and linear predictive coding (LPC). Speech intelligibility was measured in realistic acoustic noise environs by a panel of 10 subjects performing the Modified Rhyme Test. Summary data is presented along with planned future research in optimization of audio bandwidth vs bit error rate tradeoff for best speech intelligibility. 87/02/00 87N29529

UTTL: Packet voice communication
AUTH:A/NOLL, PETER; B/LEESEMANN, VOLKER; C/WESSELS, GUENTER CORP: European Space Agency, Paris (France).

ABS: The problems of transmitting packetized voice signals are reviewed and the distortions resulting from digitization, speech detection, channel errors, and constant and stochastic transmission delays are analyzed. Measures to overcome distortions, and progress and goals in experimental networks which include satellite-based experiments are discussed.
RPT# : ESA-TT-1006 DFVLR-MITT-86-05 ETN-87-90009 87/02/00 87N27867

UTTL: Robust coarticulatory modeling for continuous speech recognition
AUTH: A/SCHWARTZ, R.; B/CHOW, Y. L.; C/DUNHAM, M. O.; D/KIMBALL O.; E/KRASNER, M.; F/KUBALA, F.; G/MAKHOUL, J.; H/PRICE, P.; I/ROUCOS, S. CORP: Bolt, Beranek, and Newman, Inc., Cambridge, MA.

ABS: The purpose of this project is to perform research into algorithms for the automatic recognition of individual sounds or phonemes in continuous speech. The algorithms developed should be appropriate for understanding large-vocabulary continuous speech input and are to be made available to

the Strategic Computing Program for incorporation in a complete word recognition system. This report describes process to date in developing phonetic models that are appropriate for continuous speech recognition. In continuous speech, the acoustic realization of each phoneme depends heavily on the preceding and following phonemes: a process known as coarticulation. Thus while there are relatively few phonemes in English (on the order of fifty or so), the number of possible different accoustic realizations ds in the thousands. Therefore; to develop high-accuracy recognition algorithms, one may need to develop literally thousands of relatively distance phonetic models to represent the various phonetic context adequately. Developing a large number of models usually necessitates having a large amount of speech to provide reliable estimates of the model parameters. The major contributions of this work are the development of: (1) A simple but powerful formalism for modeling phonemes in context; (2) Robust training methods for the reliable estimation of model parameters by utilizing the available speech training data in a maximally effective way; and (3) Efficient search strategies for phonetic recognition while maintaining high recognition accuracy.
RPT# : AD-A174393 BBN-6383 86/10/00 87N18701

UTTL: A fast algorithm for the phonemic segmentation of continuous speech
AUTH: A/SMIDT, D. CORP: Kernforschungszentrum G.m.b.H., Karlsruhe (Germany, F.R.). CSS: (Inst. fuer Reaktorentwicklung.)
ABS: The method of differential learning (DL method) was applied to the fast phonemic classification of acoustic speech spectra. The method was also tested with a simple algorithm for continuous speech recognition. In every learning step of the DL method only that single pattern component which deviates most from the reference value is differences. An LPC-10 vocoder's ability to process linguistic and dialectical suprasegmental features such as intonation, rate and stress at low bit rates should be a critical issue of concern For future research.
RPT# : AD-A163307 RADC-TR-85-264 85/12/00 86N26504

UTTL: Text-dependent speaker verification using vector quantization source coding
AUTH: A/BURTON, D. K. CORP: Naval Research Lab., Washington, DC.
ABS: Several vector quantization approaches to the problem of text-dependent speaker verification are described. In each of these approaches, a source codebook is designed to represent a particular speaker

saying a particular utterance. Later, this same utterance is spoken by a speaker to be verified and is encoded in the source codebook representing the speaker whose identity was claimed. The speak is accepted if the verification utterance's quantization distortion is less than a prespecified speaker-specific threshold. The best of the approaches achieved a 0.7% false acceptance rate and a 0.6% false rejection rate on a speaker population containing 16 admissible speakers and 111 casual imposters. The approaches are described, and detailed experimental results are presented and discussed.

RPT# : AD-A161875 NRL-MR-5662 85/11/26 86N23782

UTTL: Speech recognition: Acoustic, phonetic and lexical
AUTH: A/ZUE, V. W. CORP: Massachusetts Inst. of Tech., Cambridge.
CSS: (Research Lab. of Electronics.)
ABS: Our long-term research goal is the development and implementation of speaker-independent continuous speech recognition systems. It is our conviction that proper utilization of speech-specific knowledge is essential for advanced speech recognition systems. With this in mind, we have continued to make progress on the acquisition of acoustic-phonetic and lexical knowledge. We have completed the development of a continuous digit recognition system. The system was constructed to investigate the utilization of acoustic phonetic knowledge in a speech recognition system. Some of the significant development of this study includes a soft-failure procedure for lexical access, and the discovery of a set of acoustic-phonetic features for verification. We have completed a study of the constraints provided by lexical stress on word recognition. We found that lexical stress information alone can, on the average, reduce the number of word candidates from a large dictionary by more than 80%. In conjunction with this study, we successfully developed a system that automatically determines the stress pattern of a word from the acoustic signal.

RPT# : AD-A160008 85/10/01 86N18589
used for a new rule. Several rules of this type were connected in a conjunctive or disjunctive way. Tests with a single speaker demonstrate good classification capability and a very high speed. The inclusion of automatically additional features selected according to their relevance is discussed. It is shown that there exists a correspondence between processes related to the DL method and pattern recognition in living beings with their ability for generalization and differentiation.

RPT# :KFK-4062 ISSN-0303-4003 ETN-86-98266 86/04/00 87N13635

UTTL: Automatic speech recognition for large vocabularies
AUTH: A/AKTAS, A.; B/KAEMMERER, B.; C/KUEPPER, W.;
    D/LAGGER, H. CORP: Siemens A.G., Munich (Germany, F.R.). CSS:
    (Informationstechnische Grundlagen.) Sponsored by BMFT

ABS: An isolated word recognition system for large vocabularies (1000 to
    5000 words) with 98% recognition performance was developed. It was
    implemented on an array processor for real time requirements. The speech
    signal is described by short time autocorrelation functions. Short response
    times as well as high recognition accuracies are achieved by means of a
    hierarchical classification scheme. A fast preselection stage yields a small
    number of suitable word candidates to be considered for further
    classification. To that end a linear segmentation or a segmentation based on
    acoustic or phonetic cues was performed. High selectivity is obtained by
    using fine temporal resolution and nonlinear time alignment in the final
    classification step. By taking into account phonetically identical fragments
    of words, a distinction between highly confusable words can be made.
    Speaker adaptation for new system users is performed within a relatively
    short training phase.
RPT#: BMFT-FB-DV-85-003 ISSN-0170-9011 ETN-86-97434 85/12/00
    86N31779

UTTL: A comparative analysis of whispered and normally phonated speech
    using an LPC-10 vocoder
AUTH: A/WILSON, J. B.; B/MOSKO, J. D. CORP: Rome Air
    Development Center, Griffiss AFB, NY.

ABS: The determination of the performance of an LPC-10 vocoder in the
    processing of adult male and female whispered and normally phonated
    connected speech was the focus of this study. The LPC-10 vocoder's analysis
    of whispered speech compared quite favorably with similar studies which
    used sound spectrographic processing techniques. Shifting from phonated
    speech to whispered speech caused a substantial increase in the phonomic
    formant frequencies and formant bandwidths for both male and female
    speakers. The data from this study showed no evidence that the LPC-10
    vocoder's ability to process voices with pitch extremes and quality extremes
    was limited in any significant manner. A comparison of the unprocssed
    natural vowel waveforms and qualities with the synthesized vowel waveforms
    and qualities revealed almost imperceptible

UTTL: An adaptive approach to a 2.4 kb/s LPC speech coding system
AUTH: A/YARLAGADDA, R.; B/SOLDAN, 0. L.; C/PREUSS, R. D.;

D/HOV, L. D. 0., C.S. CORP: Oklahoma State Univ., Stillwater. CSS: (School of Electrical and Computer Engineering.)

ABS: The goal of this research was to investigate (1) Adaptive estimation methods for noise suppression and performance enhancement of Narrowband Coding Systems for speech signals and (2) Autoregressive spectral estimation in noisy signals for speech analysis applications. Various prefiltering techniques for improving linear predictive coding systems were investigated. Filter coefficients were varied to optimize each filter technique to remove noise from the speech signal. A new prefilter consisting of an adaptive digital predictor (ADP) with pitch period delay was developed and evaluated. The one-dimension filter approach was expanded upon to use a two-dimensional approach to suppress noise in the short time fourier transform domain. The two dimensional approach was found to have significant potential. Fast algorithms for efficient solution of the linear estimation problem and a new recursive linear estimator suitable for rapid estimation of a signal in noise were developed.

RPT#: AD-A160312 RADC-TR-85-45 85/07/00 86N17598

UTTL: Comparison of continuous speech, discrete speech, and keyboard input to an interactive warfare simulation in various C3 environments

AUTH:A/MANSON, R. B.; B/WRIGHT, M. E. CORP: Naval Postgraduate School, Monterey, CA.

ABS: This thesis describes an experiment conducted at the Naval Postgraduate School during the period 30 October 1984 through 30 November 1984. Specifically, the experiment compares the use of continuous speech recognition equipment, discrete speech recognition equipment, and keyboard to input commands in a command and control environment. This was accomplished by using the Naval Warfare Interactive Simulation System (NWISS) as a vehicle to pose military problems to subject in a variety of light and noise environments. Although the results are not conclusive, they do show a definite advantage in using continuous speech or keyboard entry modes over discrete speech modes. Continuous speech and keyboard methods were superior in all environmental conditions.

RPT#: AD-A156830 AD-E301723 85/03/00 86N12487

UTTL: Survey of narrow band vocoder technology

AUTH:A/MCMINN, W. B., JR. CORP: Air Force Inst. of Tech., Wright-Patterson AFB, OH.

ABS: The USAF has a need to identify a vocoder to insert into a   Low Probability of Intercept (LPI) communications system. It should be small, lightweight, low power, capable of operating in many types of aircraft, and

capable of processing intelligible, natural sounding .speech at 400 to 600 bits/seconds. Two separate units are needed: one to be used in a near-term brassboard test system and one to be used in a far-term production system. Weighted characteristic values are combined through a mapping and summing procedure to form a Figure of Merit for each system. Using these characteristic values, primary vocoder candidates have been identified and are discussed in this paper.
RPT#: AD-A151919 AFIT/CI/NR-85-24T 84/12/00 85N27114

UTTL: Automatic speech recognition in severe environments CORP: National Academy of Sciences - National Research Council, Washington, DC.
ABS: Human-machine interaction by voice was analyzed. The potential for improving the safety and effectiveness of its forces by making electronic and electromechanical devices directly responsive to the human voice and able to respond by voice is recognized. The benefits of this capability are noteworthy in situations where the individual is engaged in such hands/eyes-busy tasks as flying an airplane or operating a tank. Voice control of navigational displays, information files, and weapons systems could relieve the information loads on visual and manual channels.
RPT#: PB85-121697 84/00/00 85N21490

UTTL: Man-machine communication research for robotics reported
AUTH: A/TEMPELHOF, K. H.; B/MEYER, R. CORP: Joint Publications Research Service, Arlington, VA. In its East Europe Rept.: Sci. and Technol. (JPRS-ESA-84-046) p 1-3 (SEE N85-17176 08-31)
ABS: Speech recognition systems in robotics are reviewed. Future trends in speech communication processes are examined along with primary applications. 84/12/26 85N17177

UTTL: Speech recognition: Acoustic phonetic and lexical knowledge representation
AUTH:A/ZUE, V. W. CORP: Massachusetts Inst. of Tech., Cambridge.
CSS: (Research Lab. of Electronics.)
ABS: The purpose of this program is to develop a speech data base facility under which the acoustic characteristics of speech sounds in various contexts can be studied conveniently; investigate the phonological properties of a large lexicon of, say 10,000 words and determine to what extent the phonotactic constraints can be utilized in speech recognition; study the acoustic cues that are used to mark work boundaries; develop a test bed in the form of a large-vocabulary, IWR system to study the

interactions of acoustic, phonetic and lexical knowledge; and develop a limited continuous speech recognition system with the goal of recognizing any English word from its spelling in order to assess the interactions of higher-level knowledge sources.

RPT# :AD-A137697 84/02/01 84N20742

UTTL: Applications of artificial intelligence in voice recognition systems in micro-computers

AUTH:A/CALCATERRA, F. S. CORP: Naval Postgraduate School, Monterey, CA. CSS: (Dept. of Systems Technology.)

ABS: This research investigates the use of inexpensive voice recognition systems hosted by micro-computers. The specific intent was to demonstrate a measurable and statistically significant improvement in the performance of relatively unsophisticated voice recognizers through the application of artificial intelligence algorithms to the recognition software. Two different artificial intelligence algorithms were studied, each with differing levels of sophistication. Results showed that artificial intelligence can increase recognizer system reliability. The degree of improvement in correct recognition percentage varied with the amount of sophistication in the artificial intelligence algorithm.

RPT# :AD-A115735 82/03/00 82N34132

UTTL: The effects of microphones and facemasks on LPC vocoder performance

AUTH:A/SINGER, E. CORP: Massachusetts Inst. of Tech., Cambridge.

ABS: The effects of oxygen facemasks and noise cancelling microphones on LPC vocoder performance were analyzed and evaluated. Likely sources of potential vocoder performance degradation included the non-ideal frequency response characteristics of the microphone, the acoustic alterations of the speech waveform due to the addition of the facemask cavity, and the presence of breath noise imposed by the close-talking requirement. It is shown that the presence of the facemask produces a vowel-dependent reduction in the bandwidths of the upper speech formants. In addition, the low frequency emphasis normally associated with small enclosures is shown to occur when a pressure microphone is employed for transduction. Noise cancelling microphones, which are sensitive to the pressure gradient, do not exhibit this effect. Finally, an acoustic tube model of the vocal tract and facemask is presented which predicts the absence of spurious resonances within the frequency band of typical narrowband vocoders. Evidence supporting these assertions is presented based on observed vowel spectra. Evaluations performed using Diagnostic Rhyme

Tests indicate that the presence of the oxygen facemask and noise cancelling microphone does not result in a significant increase in the LPC vocoder processing loss.

RPT#: AD-A107908 TR-584 ESD-TR-81-277   81/09/25   82N16744

INDEX

A

Accoustic phonetic approach, 111, 112
Adaptive differential pulse code modulation(ADPCM), 78, 166
Adaptive postfiltering, 92
Air/ground communications, 4
Air operations, 4
Allophones, 60
a - index, 134
Alphabet, 58
Analysis-by-Synthesis Excitation Coding, 85
Analysis-synthesis telephony, 22, 29, 86
Anomalous sentences, 131
APCM, 24
Articulation index, 36, 135
Articulation score, 33
Asynchronous Transfer Mode, 20, 37, 176
Audio manipulation, 199
Audio signal analysis, 190
Auditory masking, 56
Authentication, 12
Automatic gisting, 189, 202
Automatic speech recognition, 189

B

B channels, 6
Band-passing limiting, 137
Beam-steered microphones, 108
Broad-band ISDN, 20
Burst switching, 10

C

Carrier phrase, 133

CCIT Study Groups, 163
CELP, 183
Cepstrum, 191
Channel vocoding, 25
Cockpit Engineering, 15
Codebooks, 83, 90
Coding delay, 94
Communications networks, 4
Confusion matrices, 139
Connected word recognition, 120
Connected word recognizers, 122
Consonant recognition test, 34
Consonants, 46
Continuously Variable Slope Delta Modulation (CVSD), 24, 37
Conversational machines, 110

D

D channel, 6
Dehopping/rehopping, 14
Delta modulation (DM), 24
Decibels, 55
Diagnostic Rhyme Test, 129, 181
Differential PCM (DPCM), 22
Digital Circuit Multiplication, 20, 37, 170
Digitally controlled DM (DCDM), 24
Digital Mobile Radio (DMR), 173
Digital Speech, 9
Digital speech interpolation (DSI), 172
Dipthongs, 45
Dynamic time warping, 104, 118, 122

E

Effective vocabulary capacity, 148
Electronic Warfare, 12
Excitation Coding, 85

F

Formant predictor, 80
Formant vocoder, 25
Formants, 47, 78
Forward adaption, 94
Fractional pitch prediction, 91
Freeze-out, 172
Fricatives, 46
Full multi-point mode, 172
Fundamental frequency, 49

G

Glottis, 25
Group Special Mobile (GSM), 174

H

H Channel, 9
HF radio, 4, 12
Hidden Markov Model, 104
High-fidelity voice (HFV), 37, 166
Human equivalent noise ratio, 148
HuMaNet, 108
Hybrid coders, 75
Hybrid switching, 10
Hybrid synthesisers, 31

I

Integrated Services Digital Network, 6
Intelligibility, 33, 129
Intra-aircraft (cockpit)
  communications, 4
Isolated word recognition, 118

J

Jamming, 14
JTIDS, 6

L

Larynx, 44
LD-CELP, 95
LD-VXC, 95
Linear predictive coding (LPC),
  27, 178
Logatoms, 34
Lombard effect, 143
Low-bit-rate-voice (LBRV), 166
LPC Vocoder, 76

M

Man-machine interface, 196
Mean Opinion Score, 130, 165
Mel frequency scale, 55
Message handling system, 177
Message sorting, 199
Modified Rhyme Test, 129
Motor Theory of Speech
  Perception, 57
Multi-clique mode, 172
Multipulse coding, 87
Multipulse linear predictive
  coding, 28

N

N-gram data, 106
Nasal tracts, 25
Nasalization, 25, 45
Nasal, 46
Noise interference, 137
Nonlinear Interpolative Vector
  Quantization, 97
Nonlinear Predictive Vector
  Quantization, 97
Nonsense words, 133

O

Office dictation, 124

On board processing, 14
Open System Interconnection (OSI), 6

P

Packet switching, 9
Parameter-coding, 76
Parametric representation, 114
Pattern recognition approaches, 111
Pattern recognition model, 113
PB word list, 34
Peak clipping, 137
Perceptual weighting, 94
Perceptual weighting filter, 89
Phonemes, 29, 59
Phonetic Discrimination test, 148
Phonetic integrity, 96
Phonetic Segmentation, 96
Pitch, 80
Pitch-excited vocoders, 25
Pitch period, 76
Pitch prediction, 80
Plosives, 46
Prediction gain, 77
Predictive quantization, 77
Prosody, 59
Pulse code modulation (PCM), 22

Q

Quality rating, 130
Quantisation noise, 29

R

RASTI-method, 137
Recognition Assessment by
   Manipulation of speech (RAMOS),
   148
Regular pulse excitation and long-
   term prediction LPC (RPE-LTP), 176
Reverberation, 137
Rhyme test, 129

S

Satellite, 12
SB-ADPCM, 169
Segmental intelligibility test, 144
Shannon's theory of
   communication, 29
Sonarants, 46
Speaker authentication, 201
Speaker identification, 33, 35
Spectrogram, 52
Speech communications, 3
Speech compression, 22
Speech enhancement, 190
Speech packetization, 176
Speech perception, 54
Speech production, 25
Speech reception threshold, 130
Speech recognition, 12, 31, 111,
   123
Speech synthesis, 30, 106
Speech Transmission Index, 136
Spread-spectrum AJ Modulation,
   13
Statistical pattern recognition
   approach, 112
Stochastically excited linear
   predictive coding, 27, 28
Supplementary services, 6
Supra-segmental test, 144
Switched terrestrial network, 4
Syllables, 46, 47
Synchronous, 164
Synchronous tandem encoding,
   164
Synthesis filter, 77
Synthesizers, 30

T

Talker verification, 32
Talking machine, 21
Talkspurts, 21, 172
TASI (Time Assignment Speech
  Interpolation), 21
Teleservices, 6
Text-to-speech synthesis, 30, 100

U

Unvoiced sounds, 25
User opinion tests, 35
Uvula, 44

V

Vector excitation coding, 87
Vector PCM, 84
Vector predictive coding, 84
Vector quantization (VQ), 27, 83
Vector Sum Excited Linear
  Prediction, 90
Velum, 25
VHF/UHF ground/air,air/air, 4
Visual perception, 62
Vocal cords, 44
Vocal tract, 44
Vocoder, 20, 22, 75
Voice input, 189, 196
Voice input/output systems, 196
Voice in multimedia systems, 108
Voice mail, 20
Voice recognition equipments, 18
Voice response systems, 30
Voice verification, 198
Voiced excitation, 45
Voiced sounds, 25, 45
Voiceless, 45
Vowels, 45

W

Waveform coders, 22, 75
Written style, 65